University of London
Institute of United States Studies Monographs

I

The American Environment

The American Environment

edited by
W. R. MEAD

UNIVERSITY OF LONDON
Published for the
Institute of United States Studies
THE ATHLONE PRESS 1974

Published by
THE ATHLONE PRESS
UNIVERSITY OF LONDON
at 4 Gower Street, London WC1

Distributed by Tiptree Book Services Ltd
Tiptree, Essex

U.S.A. and Canada
Humanities Press Inc, New York

© University of London 1974

0 485 12901 9

Printed in Great Britain by
WESTERN PRINTING SERVICES LTD
BRISTOL

Contents

Contents

Foreword

These three papers formed the basis for discussion at a seminar held at the Institute of United States Studies in the University of London. The theme was the American environment and some fifty teachers of geography from universities, training colleges and polytechnics attended. The seminar was initiated by Professor H. C. Allen, at that time Director of the Institute and now Professor of American Studies at the University of East Anglia. It was made possible by the generous assistance of the Cultural Affairs Office of the Embassy of the United States of America. Although I took the chair at the seminar, the initiative for the publication of the papers has come from Professor Esmond Wright, now Director of the Institute of United States Studies. I am grateful to him for his encouragement and to Dr Howell Daniels, the Secretary of the Institute, for the practical assistance that he has given me in helping this manuscript through the press.

No elaborate integration of the three papers was attempted. It was felt that it was better to give free rein to three specialists who were prepared to offer essays on the general theme of the seminar. J. Wreford Watson, Professor of Geography in the University of Edinburgh, has had a long-standing interest in the North American literary scene and his essay on the image of nature in the United States is in the mainstream of his current work on the perception of the North American environment. R. C. Estall, Reader in Geography at the London School of Economics and Political Science, has a long record of research and publication in the economic geography of the United States, especially of its industrial geography. Gerald Manners, Reader in Geography at University College London, is a specialist in resource appraisal, with particular reference to iron ore and fuel in the North American setting. Authors were asked to prepare contributions with the student of American studies in mind rather than the specialized systematist.

All three papers are concerned with changing interpretations of familiar aspects of the North American scene and with attempts to understand its realities more fully. Each paper looks at fundamentals of the American environment in a personal way. There are other common features. Firstly, central to each are the changing scales of values that are reflected in American attitudes. Secondly, each paper is to a greater or lesser extent concerned with myths. It is of the *genius loci* of North America that great myths have grown up within it. Old myths die slowly. In the United States new myths are born more easily than in many countries. Cherished misconceptions are certainly one theme common to all of the papers. Thirdly, to a greater or lesser extent, each contribution reflects that North America has had a physical environment which to its inhabitants was in former times more than it seemed and is now less than it seems. Fourthly, each paper touches upon aspects of experience resulting from the alternation of security and insecurity. As the unorganized territory of North America was organised, the limitless was limited, the boundless was bounded, the infinite was seen to be finite, a new assurance arose—albeit an assurance rooted in book-keeping and statistics, in organisation and method. But, at a time when, in theory, understanding should confirm Americans in their assurance, the people of the United States have been thrust into a new age of insecurity. Again, at a time when the material means for confirming equality of opportunity are to hand, inequalities are sharpened. One of the conclusions common to the three contributors is that these ironies prevail because conceptual inadequacies constrain the understanding of reality. A second common conclusion is that North American man owes a debt to nature and, by paying it, he is likely to resolve at least some problems of his human nature. A third conclusion lifts familiar determinist theses to a new plane by arguing that states of mind can change the state of the nation.

W. R. MEAD

The Image of Nature in America

J. WREFORD WATSON

The image of nature in America has profoundly affected man's relation to the land and his use or misuse of its resources. The American view of nature is, of course, not far removed from a European one, especially from that of Europeans who, like Americans, had to break new land. However, the fact that Americans had more land to break and did so on a much bigger scale and at a faster pace made their reactions to nature more immediate and more compelling. These reactions changed both with time and space. Americans started off with rather a negative attitude that led them to destroy and clear away nature with ruthless efficiency, especially as the frontier advanced to the West, but subsequently they became staunch proponents of conservation and of a life tied more and more to nature.

In colonial days, as Arthur Miller points out in *The Crucible*, the 'edge of the wilderness was close by'. It was an area full of mystery and of temptation, which called out the Old Adam in people. The unredeemed lived there. Salem Man 'believed that the virgin forest was the Devil's last preserve'. Though the Puritans may have held in their hands 'the candle that would light the world', they trembled lest the forest's breath should blow the light out. The Reverend Parris found his daughter and his niece 'dancing like heathen in the forest' and he feared for the consequences: would they be put to death?[1]

How unnatural this seems today when, as Saul Bellow claims, harassed by the 'multiple excitements' of the city, men seek out the country, 'wanting to creep into hiding, like an animal'. One of his most recent protagonists, Moses Herzog, returns to nature, time and again; indeed, whenever Herzog becomes involved in another crisis, he has to 'put out this murky fire' inside him by a flight to the New England shore and 'the therapy of cold water', where he can think 'better, clearer thoughts'.[2] Megalopolitan man finds in the wilderness redemption and truth.

The Reverend Parris and Moses Herzog represent the beginning and the end of the scale of values put upon the wild. Whereas the one set wildness against the ordered world, the other looked to wildness to bring an ordered world back. This represents the profound difference time has made to the American's image of nature.

Although Miller had to be very sure that his image of the colonial view was correct, because his whole drama grows out of it, there was certainly enough evidence available. Not long after the Pilgrim Fathers had founded Plymouth they were plagued by Merrie-Mount, the rival settlement where all who wanted to dance like the heathen congregated; there a great tree was cut from the forest, and settlers, turning it into a maypole, 'revived and celebrated the feasts of the Roman Goddes Flora, or the beasly practieses of the madd Bacchinalians'.[3] Here was not just the jealousy and annoyance of a well-ordered place for a disorderly one, it was the fear of a wildness that went with the wild itself; with nature in the wild.

Perhaps the best statement of the Puritan view of nature was given in Thomas Hooker's *The Application of Redemption*. Nature, as God created it, was good and was 'made to guide and direct us in the way we should walk'. More than that, nature 'came to the full' in man. 'All creatures should have served God in man, and been happy by or through him.' But through sin man disrupted the divine order. Man thus came to 'pervert the work of the creature', and, as a result of his sin, the 'Heavens deny their influence, the Earth her strength, the Corn her nourishment; thank sin for that'. Consequently, nature fell into league with man to frustrate the gospel and put fallen instincts before the divine order. Thus nature, having been 'ruinated and laid wast' by man, could snare and ruin in return.[4]

As R. W. Harris has shown, the period from the sixteenth to the eighteenth century was crucial in the changing views of man and nature. The older view that 'there was a great universal order where Nature became a reflection of the mind of the Deity' was challenged by 'man, endowed with freedom of choice, becoming the "moulder of himself"', capable of degenerating to the level of the beasts or, by the use of his reason, becoming a heavenly being'.[5] Confusion reigned. Sir Thomas More defined 'vertue to be life, ordered according to nature', but Fulke Greville, Lord Brooke, complained

> Oh wearisome condition of Humanity!
> Borne under one Law, to another bound:
> Vainely begot, and yet forbidden vanity
> Created sick, commanded to be sound.
> Nature herselfe, doth her own selfe defloure...

Such confusion certainly obtained in America where it was confounded by what Locke called the 'enthusiasm' of the Puritans, who so frequently put their revelation about nature before their native reason. In any event the ambiguity of the times did allow those who wanted to believe in a nature cracked, if not crooked, to have their say. And their say, at any rate in New England, often held sway.

It is not surprising, therefore, that the sins of those New Englanders, Hester Prynne and Arthur Dimmesdale, in Hawthorne's *The Scarlet Letter* were committed in the woods. Might they not have been committed *because* of the woods? Hawthorne did not believe this; in fact he was to show that, in the end, they might not really be sins, and that nature, though kernel to the Fall, was also the seed of resurrection. None the less, Salem Man took nature to be sinful, or at any rate the occasion if not the source of man's downfall.

Hester Prynne was well aware of the temptations of the forest. Her life in town was grim in the extreme. Caught in adultery and about to have a child by a man whose name she would not reveal, she had been put in the stocks, and thereafter had to live at the very extremity of the settlement. But this was the edge of the wilderness where her love began. She both shunned and was drawn to the wild. Hawthorne tells us that 'the primeval forest...to Hester's mind, imaged not amiss the moral wilderness in which she had so long been wandering'. On the other hand, she felt that her 'intellect and heart had their home, as it were, in desert places, where she roamed as freely as the wild Indian in his woods'. Hester was divided in her reactions. Hearing that her long-absent husband, whom she had thought dead, was intent on avenging himself on her lover, Arthur Dimmesdale, she seeks to warn the young man. After meeting in the forest, they decide to run away together. Suddenly the forest becomes their way out. In spite of her fear of it, Hester wants to give way to it. She cries: 'Is the world, then, so narrow?...Doth the universe lie within the compass of yonder town? Whither leads yonder forest track? Backwards to the settlement, thou sayest? Yes; but onward, too. Deeper it goes, and deeper, into the wilderness...There thou art free!'[6]

The flight to the forest was a major theme of early American writers. Although Hester and her lover decided to return to the town rather than flee, many others sought refuge from their problems by evading them in the wild. This annoyed a man like President Timothy Dwight of Yale who complained bitterly of unruly elements who, 'impatient at the restraints of law and morality', fled from their responsibilities in the coastal settlements into the up-state wilds. An early historian of the Revolution, David Ramsay, told how 'many disorderly persons fled from the old (coastal) settlements to avoid the restraints of civil government'.[7] It must have seemed to the responsible, orderly, sober, and god-fearing people of early America that the wilderness went with irresponsibility, disorder, intemperance and godlessness. And there was an element of truth in this; the freedom of the forest was the freedom to fly. Instead of standing up to themselves, Americans were constantly tempted to run away from themselves. In a way, wildness was weakness. Thus, to go back to nature was to step out of grace.

It was not a far cry, perhaps, from temptation to malevolence. Nature was seen as *against* man: harsh, stern, exacting and hostile. This impression was partly a matter of experience. For example, the American climate had extremes rarely known in Britain. Killing frosts could damage growing or ripening crops. Early spring rains combined with the late melting of winter snows could sweep down from the mountains in devastating floods. The 'toile of a new Plantation' cried Edward Johnson in his *Wonder-Making Providence* is 'like the labours of Hercules never at an end'. But it was more than just a matter of toil and tears. Men believed that they were, in Hooker's words, up against a force breeding 'nothing but vanity', that 'brings forth nothing but vexation. It crooks all things so that none can straiten them.' Nature was itself full of guile and deceit and out to 'justle the law out of its place and the Lord out of His glorious Soveraignty'.[8]

Men felt no compunction, then, in turning against nature. In doing so, they could take out their Fall, as it were, by felling the world. They could help to justify themselves by undoing nature. Melville, for example, went well beyond Hawthorne. He was concerned not so much with man undone by nature as with man seeking to out-do nature. *Moby Dick* deals not with a flight *to* nature, as man seeks the wilderness to give way to the wildness in himself, but with the pursuit *of* nature, as man challenges nature itself to prove his own nature. The story is told by that strange character who, at the beginning, says 'Call me Ishmael'. We are dealing with the American as Ishmaelite, as hunter and killer, as one who, in Melville's words, had suffered enough from 'the wolfish world to want to turn his splintered heart and maddened hand' against it. Melville makes the owners of the whale-ship two Quakers, but these are Quakers with a difference. Although they would not dream of spilling human blood, they have made their fortunes out of tuns and tuns of Leviathan gore. This attitude reflected a view of nature in which men could take out their frustrated vindictiveness on the world: it was fair game for plunder. The arch contender is, of course, Captain Ahab, under whom Ishmael signs on. When Ahab sails out to sea in the *Pequod*—whose name is so reminiscent of the tribe of Pequots the Puritans had smote hip and thigh, putting even their women and children to fire and the sword—he calls his men together and tells them he is after the white whale, Moby Dick, who had once 'devoured, chewed up, and crunched' off his leg. ' "Aye, aye! and I'll chase him round Good Hope, and round the Horn, and round the Norway Maelstrom, and round perdition's flames before I give him up." ' And then he appeals to his men. ' "And this is what ye have shipped for, men! To chase that white whale...over all sides of earth, till he spouts black blood and rolls fin out. What say ye, men, will ye splice hands on it, now?"..."Aye! Aye!" shouted the harpooners and seamen, running closer to the excited old man: "A sharp eye for the White Whale; a sharp lance for Moby Dick!" '

One of the officers on board the whaler, Starbuck, tries to remonstrate. He turns on the Captain and cries out in protest '"Vengeance on a dumb brute...that simply smote thee from blindest instinct! Madness! To be enraged with a dumb thing, Captain Ahab, seems blasphemous."' Ahab, stung to fury by this and likening himself to a bound and afflicted being suffering the injustices of the world, replies '"How can the prisoner reach outside except by thrusting through the wall? To me, the white whale is that wall, shoved near to me. Sometimes I think there's naught beyond. But 'tis enough. He tasks me; he heaps me; I see in him outrageous strength, with an inscrutable malice sinewing it. That inscrutable thing is what I chiefly hate; and be the white whale agent, or be the white whale principal, I will wreak that hate upon him. Talk not to me of blasphemy, man; I'd strike the sun if it insulted me."'

This is the supreme American hero—the self against the world: the unbowed and undefeatable self against whatever hostility or savagery nature might show. Ahab was particularly American in that he rose to be a 'Khan of the plank, a King of the sea, a great lord of Leviathans' from quite humble origins, and thus justified the 'great democratic God who...ever cullest Thy selectest champions from the kingly commons'.[9] He was one of the leaders of a race of leaders, the whale captains of America, that had made their way through their own wit and grit, and skill and strength.

But they were a race of killers and as such influenced the American mind. An Ahab had entered into the people and stamped his harsh implacable image on the land. His spirit went down through the ranks, for it was not only in huge figures like Ahab that the killer-American was let loose, but in quite ordinary men everywhere. While Ahab had been ploughing the seas, the killer of whales, the ploughmen of the West had started to cut out the forest and strip off the sod, from the Appalachians to the Rockies, in the greatest ravagement of nature known to history. What it took Europe about 6,000 years to do, Americans did in 150 years. They exterminated the wild pigeon, practically eliminated the moose and the elk and virtually destroyed the buffalo. In fact they went much further than the Europeans who under strict game laws in royal and baronial forests preserved much of the continent's natural condition. Even in a country as populous as Britain some of the primitive wild-life, such as the *Bos primigenius*, has been protected until this day. By contrast, in a matter of two to three generations the killer American had altered the balance of life throughout his continent. In so doing he changed the whole landscape.

It may be argued that he needed the whale oil to light his homes, pigeons for pies, moose for meat and the skin of the buffalo to keep him warm. Wild life was as much a resource to be used as coal or iron. We are dealing with exploitive man in a world to be exploited. This was the kind of man shaped by the utilitarianism of Hume and the pragmatism of Adam Smith, a man soon to come under the influence of Darwin and Herbert Spencer. The 'struggle for existence' became the philosophy of the day on both sides of the Atlantic. The same rampant individualism that rolled out the factory in Europe unrolled the frontier in America. Most migrants who went west were poor; but they saw a chance to grow rich. American wealth could get mankind away from its crippling poverty. The struggle threw up the frontier aristocrat, who carved a fortune out of the land through his own wit and endeavour. Men could do what it was in their nature and power to do, or what it was to their help and gain to do. Yet it was not their need to exploit the world that became the problem; rather it was their desire to be paramount and unchallenged—or, simply, their greed in devouring nature.

This was part of the theme in Fenimore Cooper's novel *The Pioneers*. In it he causes Leatherstocking, the old hunter who made his livelihood from killing game, to object passionately to the indiscriminate killing practised by the pioneers who were pushing into and replacing the forest. Cooper centres his protest on a scene where the farmers engage in a riot of killing to keep the carrier pigeons off their crops. At the time the pigeon migrated 'the whole village seemed equally in motion with men, women, and children. Every species of firearms, from the French ducking-gun, with a barrel near six feet in length, to the common horseman's pistol, is to be seen in the hands of the men...', while boys armed themselves with 'bows and arrows, some made of the simple stick of a walnut sapling, and others in rude imitation of the ancient crossbows...The signs of life apparent in the village drove the alarmed birds from the direct line of their flight towards the mountains, along the sides of which they were glancing in dense masses, equally wonderful by the rapidity of their motion and their incredible numbers'. The people hurry to the edge of the valley and their 'posted themselves, and, in a few moments, the attack commenced'. On hearing the salvos of firing Leatherstocking, the hunter, is appalled at the 'wasteful and unsportsmanlike execution'. He notices that in people's madness to kill none 'pretended to collect the game, which lay scattered over the fields in such profusion as to cover the very ground with the fluttering victims'. Leatherstocking calls out in protest and argues hotly with one of the leaders, Billy Kirby. '"It's wicked to be shooting into flocks in this wasty manner"' he cries. '"If a body has a craving for pigeon's flesh, why! it's made the same as all other creatures, for man's eating; but not to kill twenty and eat one."'' To this Kirby retorts '"If you had to sow your wheat twice and three times as I have done, you wouldn't be so massyfully feeling'd towards the divils."''[10]

It was not the use but the abuse of wild-life that became Cooper's concern. Few listened to him. Twenty years after the pigeons were shot out of the sky, the buffalo began to be shot off the plain. The killer-American rode again. Oddly enough, he could be the most pleasant of persons—for instance, a man like Washington Irving. When the young reporter was out west he joined a troop of soldiers that were hunting buffalo 'for a supply of provisions'. He and his friends went simply for the sport. 'After a while', he wrote, they surrounded 'an unsuspecting herd of about forty head, bulls, cows and calves'. Mr L. was the first to draw blood. Losing ground, he took a chance, 'levelled his double-barrelled gun, and fired a long raking shot'. Since it was into a bunched mass of terrified animals it could hardly miss. In fact 'it struck a buffalo just above the loins, broke its backbone, and brought it to the ground'. This success excited Washington Irving still more. 'Galloping along parallel to the herd I singled out a buffalo, and by a fortunate shot brought it down on the spot. The ball had struck a vital part; it would not move from the place it fell, but lay there struggling in mortal agony.

'Dismounting, I now fettered my horse and advanced to contemplate my victim...Now that the excitement was over I could not but look with commiseration on the poor animal that lay struggling and bleeding at my feet...He had evidently received a mortal wound, but death might be long in coming. It would not do to leave him here to be torn piecemeal, while yet alive, by the wolves that had already snuffed his blood and were skulking and howling at a distance.' The young sport therefore despatched him with another bullet. 'While I stood meditating and moralizing over the wreck I had so wantonly produced' Irving confessed, 'I was rejoined by my fellow sportsman who...soon managed to carve out the tongue of the buffalo, and delivered it to me to bear back to the camp as a trophy.'[11] The rest of the beast was left to the wolves and the ravens.

Men behaved like lords of creation, killing when and where they liked. They would often kill for the mere fun of it. Robert Penn Warren gives a vivid flash of this in a calm but appalling scene of two Southerners taking a trip out west, engaged in a senseless slaying of wild life as they stand nonchalantly on board a Mississippi river boat and shoot duck after duck when the startled birds fly overhead. No one on the boat remonstrates at this crass slaughter, done simply to show off skill as marksmen. Penn Warren describes how 'Off Arkansas, again toward evening, a great flight of duck rose from the margin of the flooded woodlands westwards, wheeling over the river. A gentleman stood on the deck and behind him a Negro, a body servant. The gentleman would fire a fowling piece into the midst of the flight, would hand the discharged weapon back to the Negro and receive another, ready and primed, and would fire again, almost all in one motion. Another gentleman, watching this for a bit, went away and returned with a pistol. Amid the incredulous chaffing of the company, he lifted that instrument, laying the barrel across the left wrist, and fired. The duck that his ball hit seemed to hang motionless for an astonishing moment, then slithered crazily sideways down the air. "Bravo! Bravo!" someone cried, and the marksman laughed in amiable modesty, and lounged on the rail. The wounded duck, still struggling in the water, fell astern.'[12]

These gentlemen were not even out for a hunt; they simply

amused themselves, passing the time of day by proving their aim. No wonder the great flightways of nature became the blightways of man, and the wild duck disappeared in such numbers the Americans soon had to go up to Canada to shoot them! Penn Warren's story is no 'story'; men did kill off game for the fun of it. For example, even as intelligent a man as Francis Parkman admits that dreading 'the monotony and languor of the slow-moving emigré camp' (with which he was associated, during part of his trip along the Oregon Trail), he 'amused' himself killing off the wolves and other wild life he saw on the prairie.[13] On Parkman's first buffalo hunt he wounded a bull but did not kill it. He gave no second thought to what happened then: did it die of its wound, or was it, weakened by its loss of blood, attacked and torn to pieces by the wolves?

It was a callous age, bred in the Indian wars, and schooled in the war of each-against-all in the industrial revolution. Nor has the mentality of it passed. The killer-American is still on the road. Steinbeck gives us what must be one of the cruellest vignettes of the American at the kill found in America's story. And this happened in the nineteen thirties. Across the great way west of America a turtle starts to crawl, 'turning aside for nothing, dragging his high-domed shell over the grass'. Coming to the highway the turtle is challenged by a steep embankment. As this 'grew steeper and steeper, the more frantic were the efforts of the land turtle'. But at last it overcomes the barrier and crawls, determinedly, on the road. At this point 'a light truck approached, and as it came near, the driver saw the turtle and swerved to hit it. His front wheel struck the edge of the shell, flipped the turtle like a tiddly-wink, spun it like a coin, and rolled it off the highway. The truck then went back to its course along the right side'.[14] Here is an ordinary American, a truck-driver, about his ordinary business, carrying goods west along the oldest American trail, who suddenly, on seeing a last vestige of wild-life, a little land turtle, feels all the killer-instinct in him aroused, swerves out of his way to hit it, before swinging back, indifferently, to the road. This is sheer uncalled-for destructiveness, the end product of hundreds of years of Americans swerving out of their way to give vent to the destructiveness in their system.

Violence has become in-built into the American character. As Rap Brown, the Negro activist, once claimed, 'Violence is as American as cherry pie!' Americans grow up on it. In Arthur Schlesinger's tract on *Violence: America in the Sixties* he writes: 'We must recognize that the destructive impulse is in us, and that it springs from some dark intolerable tension in our history. No nation, however righteous its professions, could act as we did without burying deep in itself—in its customs, its institutions, its conditioned reflexes and its psyche—a propensity towards violence.' The widespread and often indiscriminate killing of wild life, and, with it, the continued attack on the Indian, instilled in America a killer-instinct that had far-reaching consequences. It breaks out in religious intolerance, industrial strife, race riots and war, both at home and abroad. Schlesinger continues: 'It is almost as if this initial experience (of attacking the Indian and killing off the wild life on which he depended) fixed a primal curse on our nation—a curse which still shadows our life'.[15]

This curse went back to the very blessing that America cherished most—its freedom. America's confrontation with nature occurred at a time when the idea of personal liberty was strongest. 'Give me freedom or give me death' was the cry of the revolution. Unfortunately, freedom for the individual often meant death to the land. The freedom to be an American was an almost unlimited right to make free with America. The break-through to the West was especially catastrophic, for by then man's power to destroy had become enormous. The frontier may have made a great contribution to the character of the American people, but it was a tragedy for the American land. Too often attention has been paid to the people, rather than to their impact on the land. For F. J. Turner the frontier meant 'self-reliance, individualism, democracy, and self government'. But the frontier also meant deserts of erosion in Appalachia, and cut-over barrens in the Ozarks; it loosed new floods on the Mississippi, and set dust storms never before seen blowing across the Great Plains. In Fenimore Cooper's novel *The Prairie*, he cries out against those pioneers 'the rank smell' of whose passing-plunder 'showed the madness of their waste'.

The history of frontier freedom was a frontier geography of

waste and ruin. The whole basis of existence degenerated. America, in Vance Packard's words, became a nation of waste-makers. G. T. Renner puts it very dramatically. Writing in the nineteen-thirties he said: 'After a mere 150 years of American existence, some 85 per cent of our wild game is gone, 80 per cent of our timber has been cut, about 67 per cent of our petroleum reserves, 65 per cent of our lead and zinc, 60 per cent of our high-grade iron, and 52 per cent of our copper have been used up, while at least 10 per cent of our cultivable land has been ruined beyond repair. Yet 150 years represent only the life-time of two men—what a disastrous pair of lifetimes'.[16]

Although this is true, there have been some people who have cared. Steinbeck was fair enough to point this out. Along the same way west that had produced his killer-American came another driver, a woman who also saw the turtle mounting the road and 'swung to the right, off the highway, the wheels screamed and a cloud of dust boiled up. Two wheels lifted for a moment and then settled. The car skidded back on to the road, and went on, but more slowly'.[17] Such an attitude is also part of the American myth. Destruction has never had all its own way. Care for life is also instinctive and conservation has grown out of it. Although weak to begin with, conservation has grown in strength and now matches destruction in every scene. From the beginning many Americans have respected and cared about nature. Even the Puritans were by no means all suspicious. Jonathan Edwards, for example, frequently acknowledged his debt to nature. As a minister, he often got inspiration from nature, catching 'a calm, sweet cast or appearance of divine glory in almost everything'. He was Puritan enough to claim that salvation was by grace alone, made known 'through a sense of divine excellency in our hearts'; nevertheless, its truth could be confirmed by nature, 'in a notion of it in our heads'.[18] His keen interest in science allowed him to accept a nature that, in the words of Basil Willey, 'could fuse harmoniously with the presuppositions inherited from Christianity. For what had science revealed? Everywhere design, order, and law...'[19]

However, the world was not to be loved for its own sake; men had to remember that 'It was never designed by God to be our home'. But if it was treated as though it were 'a place of preparation' for the heavenly world, then it could certainly serve God's purpose. This idea that men should not set their heart upon the land but ought, none the less, to use it for higher and more lasting ends was a compromise that became typical of many devout Americans. While they accepted the vanity of this world, they also undertook the stewardship of their possessions and, to this extent, they helped to conserve nature. They became true husbandmen of their resources.

Of more importance than the mere husbandry of nature was the belief in nature itself, a belief that led many Americans to identify themselves with nature. There were, for example, the forest-men, immortalized in Fenimore Cooper's Leatherstocking tales. Such men did not stop to argue out the good or the bad in nature but, taking the rough with the smooth, gave themselves up to the wilderness. Yet, as Cooper points out, the wilderness did not necessarily bring out the wild in them: on the contrary it made them more manly, helping them to get away from what was contrived and artificial, to a natural disposition based on honesty and truth. They lived in a world that 'came fresh from the hands of their Almighty Creator'.

Men like Leatherstocking who acted as though Nature rather than Society had the rights of the matter made it easier for the new concept of nature to arise, an attitude that is best expressed in Emerson's essays. Emerson opposed the old Puritanism whose 'devotee flouts nature', and claimed that 'In the woods we return to reason and faith'. Emerson was concerned at the loss of contact with nature. By his day, much forest had been cut away in the East and men seemed far removed from the wild. Most of them were living in a man-made environment of brick and clap-board, and a great many worked in entirely artificial circumstances. The fear was no longer the outer wilderness but the inner desert—the barrenness and aridity of a life run on impersonal and often unnatural lines. Men were caught up by profit and loss, organization and method, blueprints and machines to such an extent that they had lost touch with the living realities.

Emerson urged men to return to nature and find there a source of transcendental inspiration. He felt men should get away from 'the artificial and curtailed life of the cities'. He claimed that 'A life in harmony with nature will purge the eyes to understand'. This was because 'the noblest ministry of nature is to stand as the apparition of God'. Such was quite the reverse of the Puritan view and, as it became more and more accepted, gradually gave Americans a changed attitude to nature. Emerson's attack on people whose outlook was 'to put an affront upon nature' led to a revolution in mind that subsequently brought about a transformation in the land. Americans began to preserve the wilderness, before it disappeared from their midst.

Thoreau went further than Emerson by claiming that unless men made a place for nature their own natures would suffer. 'In wildness' said Thoreau 'is our preservation'. Now Thoreau was no romantic. He recognized that to stay and work in the country could be soul-destroying where 'the better part of a man is ploughed into the soil for compost'. Nevertheless men had to be broken from 'having no time to be anything but a machine'. He confessed 'I went to the woods...to live deliberately, to front only the essential facts...to live deep and suck out all the marrow of life'.[20] Few could do as he did and take years out to find in the wisdom of the woods the meaning of their days. Nevertheless, many followed him to the extent that they sought out some wild spot, put up a cabin, and took their family into retreat with them for an annual holiday. To this day, millions of Americans stream out of their cities every summer and try, by going into the woods, once again to recapture something of the simplicity and refreshment found at the hand of Nature. Here they wear the fewest clothes, live in a log hut, chop their own wood, cook over an open fire, fish in stream or lake, and walk along nature trails, looking for loon's nest or beaver's lodge.

As a result, woods are at a premium. There is today a desperate attempt to preserve them. People buy up sections of farm wood-lots or, in the case of run-down places, take the whole farm over and replant it. Developers stake out claims to big blocks of country lost in wooded land and then sell the lots for cottage colonies. Water-line lots are particularly valuable. Second-growth scrub beside rivers or lakes where logging companies have cut out the timber beyond replacement, will sell as well as city property, even if logging roads are the only access routes. A whole new geography has sprung up, centred on the restoration of the tree after the frontier onslaught had virtually destroyed the forest. In helping to re-create man, nature was itself recreated.

One of the best examples of this situation lies in Marquand's *The Late George Apley*. This story starts about the end of Thoreau's generation and continues into the nineteen-twenties of the present century. It thus covers a period in which the family camp, deep in the woods, was at the height of its influence and profoundly affected both the people and the land. It tells of a wealthy family in Boston who found one of their most constant sources of inspiration at a camp in Maine. To this they returned through three or four generations, since they felt it kept them down to earth, reliant, adaptive, sane and free.

The camp originated from a canoe trip one September when the maples were a vibrant red and the birch a rich gold. 'We were weary indeed', George Apley recounts 'as the canoes neared the golden bow of beach that fringed the wooded slopes of Pequod Island, but it was a carefree, happy weariness. As we went for a dip in the lake...we were alone in a wilderness of woods and waters, alone save for the mournfull call of the loons and the splash of an occasional fish.'[21] They were completely away from traffic and business and the whole façade of the city, and at once George Apley decided to buy the island as his refuge and strength.

It is interesting that the ship, Pequod, in Melville's *Moby Dick* was the vehicle of vengeance, in which man hunted down and tried to destroy Nature, whereas the island, Pequod, in Marquand's *The Late George Apley* was a refuge of renewal, where Nature was sought to help man redeem himself. The use of the Pequod symbol in these very different ways shows how differently men viewed nature as time passed.

George Apley is but one example of the millions of individuals in America who have staked their claim in the wilderness as a means of regenerating themselves. In this way they have helped to regenerate the wild as a vital part of the American landscape. The circle has come full round. Apley, a descendant of those Puritans who had at one time abjured nature, now felt it was his duty to retire there, year after year, to get into the heart of the wilderness. Boston, which had once hacked away at the trees, now felt the trees were part of a heritage that had to be maintained. This change in mind changed the whole land and kept an ever growing amount of Maine and Vermont, of the Green Mountains and the Berkshires, in renewed beauty that might otherwise have ended up mere cut-over barrens.

The wilderness began to be looked on not only as a renewing but as a redeeming force. How very different from Salem days when the forest was 'Satan's last preserve'. There is, in fact, a complete turn-about in the morality of the woods. In *The Scarlet Letter* the forest had spelt temptation. When Mistress Hibbins saw Hester Prynne and Arthur Dimmesdale together in the woods, she could only think one thing of them—and that was *sin*. They had gone back to the forest to resume their illicit love. By contrast when people saw George Apley, a married man, take walks in the woods of a Saturday morning with Clara Goodrich, a married woman, they did not raise an eyebrow. The couple were in the *woods* together! Though side by side, on their own, hidden from the world, they were *nature lovers*: this explained and excused everything. George Apley's wife, Catharine, herself accepted the situation. Marquand assures us that 'she always understood completely his purpose in taking these weekly journeys' even though they were spent with another man's wife. This was because, as her husband wrote in his diary, 'I always feel better after searching for birds with Clara. The world seems to revolve more easily. I return home in a better mood'.[22] In the woods, then, all was well. Here America could only find inspiration. By the twentieth century the jungles where sin stalked had become the cities. Hence Upton Sinclair's title for his grim novel of Chicago life—*The Jungle*.

Geographers must be ever watchful for these trends, for changes in states of mind can change the state of the country. With this new attitude to nature, nature could take on a different role in the land. It was seen to be as important to the lives of people—and certainly to their inner lives—as the city. The country came to reduce the city's wrongs. If given the chance, nature could restore a balanced world and help to build up a whole environment. It could in this way support and strengthen America's ideal, as expressed by Emerson, of achieving the Whole Man. Few things were of greater importance. Hence the nationwide effort to improve the countryside and gain back from it the renewal many people had lost. To help heal the country could, in turn, make a healer-American who would counter-balance what the killer-American had done.

More and more voices were raised on behalf of nature. The healing instincts of the healer-American were roused by stories of waste and destruction. The American freedom to make free with the land had to be changed to an American duty to give the land its due. Perhaps the first person who consciously strove to this end was George P. Marsh, often called the Father of American Conservation. In 1864 he published his book on *Man and Nature*, in which he tried to measure 'the extent of the changes produced by human action in the physical conditions of the globe we inhabit' and made a strong plea for 'the restoration of disturbed harmonies and the material improvement of waste and exhausted regions'. Scarcely ten years later, Hough, the superintendent of the U.S. Census, wrote an influential paper on *The Duty of Governments in the Preservation of Forests*. The mental tide was turning. By 1891 the President was empowered to set up forest reserves; in 1901 the Government widened its interest in forestry by creating the Bureau of Forestry, later known as the Forest Service. Forests were at first preserved simply to prevent them from falling into the hands of unscrupulous or unthinking entrepreneurs. As early as 1872, however, a more positive approach had been made when the forest at Yellowstone was turned into a park 'for the benefit and enjoyment of the people'. Not all Americans could afford to buy their own cottage in the wood, but through State and National Parks they could get out by train and by car into countryside publicly held for their convenience. A National Park Service was created in 1916. There are now about 50 national parks totalling over 22 million acres.

Most of these have, however, become 'citified' to a certain extent; they have their lodges, motels, cafes, tourist shops, caravan sites and tenting places, all marked out and fenced off, with flush toilets, wash places, showers and even cooking facilities provided. Cars, car parks, and gasolene stations are everywhere in evidence. Consequently, while many people get into the country, not so many become part of it. George Apley found this. 'I had thought' he wrote to a friend, 'that on first coming to Pequod Island we might get away from things. I suppose that this was rather too much to hope for'. The retreat had become a rendezvous. Cottages were built around the main lodge, and guests were invited up for the summer. Even guest lecturers were brought in, to elevate people's minds with lectures on poetry, philosophy, music and art. 'It sometimes seems to me' Apley complained 'that Boston has come to Pequod Island.'[23]

For many Americans, Boston had indeed come to Pequod. Their own little camp in the wood had become citified, and the State and National Parks were full of the latest city conveniences at trailer and camping sites. Consequently, the idea began to arise of preserving truly wild areas into which neither train nor car would be allowed, but where people would have to walk, ride on horseback, or canoe to get in or out, and thus shed the appurtenances of civilization and really get close to nature. The wilderness had once been thought of as bringing out the wildness in man, of leading to the dances of Merriemount and the witch-parties of Salem. Now the wilderness was considered as a last chance for man to save his own nature. It was in this light that the Wilderness Act was passed in 1964 to provide truly wild areas where men, in getting to grips with nature, could also conserve what was natural in themselves as well.

It is interesting that in Bellow's story about Herzog his hero seeks out the countryside at every major crisis of his life. It is as if Herzog feels he can only get to grips with himself when he is in league with nature. The thing that most appalls him is how man is making a mere *business* out of life. He writes: 'Dear Mr President, Internal Revenue regulations will turn us into a nation of book-keepers. The life of every citizen is becoming a business. This, it seems to me, is one of the worst interpretations of the meaning of human life history has ever seen. Man's life is not a business.'

If business we must have, let it be the business of living. And this must certainly include what is the business of nature. Here Americans have done not only a lot for themselves but a great deal for the world. In spite of the ruthless and often destructive attitude many of them have taken towards nature (in which however they were very little different from other colonizers of their age), they have gradually built up a strong positive position where they have taken the lead in world conservation. In so doing they have brought men everywhere to realise the debt that is owed to nature, the payment of which can profoundly affect the nature of man.

NOTES

1. Arthur Miller, *The Crucible*, Secker & Warburg, London, 1966, pp. 5, 10.
2. Saul Bellow, *Herzog*, Weidenfeld & Nicolson, London, 1965, p. 27.
3. William Bradford, *History of Plymouth Plantation*, quoted in Perry Miller and T. H. Johnson (eds.), *The Puritans*, Harper, New York, 1965, i, 108.
4. Ibid., pp. 294, 296, 293.
5. Harris, R. W., *Reason and Nature in Eighteenth-Century Thought*, Blandford, London, 1968, pp. 22–3.
6. Nathanial Hawthorne, *The Scarlet Letter*, Collier-Macmillan, New York, 1962, pp. 179, 193, 191.
7. R. J. Hooker (ed.), *The American Revolution: the Search for Meaning*, Wiley, New York, 1970, p. 21.
8. *The Puritans*, i, pp. 155, 296, 294.
9. Herman Melville, *Moby Dick*, Signet, New York, 1960, pp. 66, 85, 166, 167, 124.
10. Fenimore Cooper, *The Leatherstocking Saga*, Collins, London, 1955, pp. 679, 681.
11. Washington Irving, *A Tour on the Prairies*, Harlow Publishing Co., Oklahoma City, 1955, pp. 142–3.
12. Penn Warren, *Band of Angels*, Eyre and Spottiswoode, London, 1956, p. 86.
13. Francis Parkman, *The Oregon Trail*, O.U.P., London, 1944, p. 54.
14. John Steinbeck, *The Grapes of Wrath*, Heinemann, London, 1960, pp. 12–13.
15. Arthur Schlesinger, *Violence: America in the Sixties*, Signet, New York, 1968, p. 31.
16. Renner, G. T., *Conservation of National Resources, an educational approach to the problem*, Wiley, New York, 1942.
17. *The Grapes of Wrath*, p. 13.
18. O. E. Winslow (ed.), *Jonathan Edwards: Basic Writings*, Signet, New York, 1967, pp. 85, 131.
19. Basil Willey, *The Eighteenth-Century Background*, Penguin, Harmondsworth, 1962, p. 12.
20. Henry Thoreau, *Walden and Civil Disobedience*, Norton, New York, 1966, p. 61.
21. John P. Marquand, *The Late George Apley*, Little Brown, Boston, 1938, p. 170.

22. Ibid., p. 199.
23. Ibid., p. 175.

Since this essay was prepared, several American books with a bearing on the subject have been published, viz:

Daniels, P., *The Stewardship of the Land: a selected bibliography of current readings*, NYS Office of Planning Services, New York, 1973, 16pp.

Hirth, H., *Nature and the American: three centuries of changing attitudes*, University of Nebraska Press, Lincoln, 1972, 250pp.

Kline, M.B., *Beyond the land itself: views of nature in Canada and the United States.* Harvard University Press, 1972.

Opie, J. (ed.), *Americans and environment: the controversy over ecology.* Heath, New York, 1971. Contains a good survey of early American views on nature.

Population and Environment

R. C. ESTALL

'But what are you going to do for me? I'm no Negro. I'm no slave. I'm a free white carpenter. What are you going to do for me?'

In a flash came the answer. 'We will give you a farm. Uncle Sam has a farm for every one of us.'

In this arresting style J. Russell Smith and M. Ogden Phillips opened the second edition of their classic geography of *North America*.[1] The question and the historic reply were first heard at an election rally held by Abraham Lincoln shortly before the Civil War. And the astounding thing was that such a promise could be kept. Under the Homestead Act of 1862 and its later amendations some 290 million acres were to be settled by homesteaders, an area approximately nine times that of England and Wales. From the foundations of the Union to the present time more than 1,100 million acres of Federal land have been disposed of to various users by sale or grant. Here, then, in the early 1860s was a nation of but 31 million people, blessed with apparently limitless resources, a major obstacle to the realization of which was a shortage of population. Only a decade or two before the Homestead Act was passed, wise men were maintaining that it would take 500 years or more to settle this almost limitless land. In this context, people were undoubtedly a boon. More people would mean an expanding labour force, create a larger and more varied structure of market demand, permit economies of scale to be realized and provide a widening range of job opportunities. And so it was to prove. Between 1860 and 1900 about 14 million immigrants were welcomed and absorbed. Total population rose by 45 million. Yet resources remained apparently plentiful, living standards rose higher and higher, the quantity of productive land space per head remained at levels which no other advanced nation of the time could match, and some 700 million acres of land remained in the Federal Domain. Unusual, and unusually bold, was he who questioned the patterns of development, the profligate use and careless destruction of resources. Abundance, as Smith and Phillips said 'is the most important key to American history [both] before Abraham Lincoln...and for many years thereafter'.[2] The philosophy of plenty was clearly reflected in Lincoln's promise, and can be seen at work until very recent times in American attitudes both to population growth and the use of resources. Such attitudes were reflected also in a calm acceptance of or a complete indifference to environmental damage, especially where this was caused in the pursuit of 'growth'.

On 20 November 1967, just over a century after Lincoln had made his famous promise, President Johnson and his retinue gathered under the National Census Clock in Washington D.C. as it ticked off the growing population, awaiting the arrival of the 200 millionth American. But by this time attitudes to population growth and its implications were becoming rather ambivalent, and the celebration of the event was somewhat muted. Expressions of pride and pleasure, and the anticipation of further economic growth were to a degree soured by the now widely publicized views of conservationists and 'environmentalists' (a rather newer breed). The old attitude of mind had paid little heed to the problems accompanying large population expansion and rapid changes both in patterns of living and in geographical distribution. In the late 1950s and early 1960s, however, concern about these matters began to grow, and the assumed arrival of the 200 millionth American, coupled with very high concurrent projections of population growth for the remainder of the century, heralded a sizeable campaign concerned with the threat, so-called, of 'overpopulation'. Stewart Udall, then Secretary of the Interior, writing a Foreword to his Department's report *The Population Challenge* stated flatly that 'The greatest threat to quality living in this country is overpopulation'.[3] He spoke of the challenge of a soaring population, the shrinking amount of space per head, and the gathering storm of conflict on space allocation, on resource utilisation and on the preservation of the quality of the environment. The sub-headings of various sections of the report—'A Rationed Tomorrow', 'So Much and No More', 'Planning for the Pinch' —all reflect urgent concern about the size and rate of growth of the population of the U.S.A. with its increasing pressure on land and other resources. The report rejected out of hand the more sanguine attitudes of other sections of American opinion as based on the narrow economic pre-occupation of 'commodity dealers and chambers of commerce'.

The differing view was not entirely confined to those with narrow business interests. A sizeable number still believed, on more general grounds, that population growth posed no serious problem for the future. In some respects the 'environmentalists' overstated the case, especially where it related to space standards. In this matter, emphasis could easily be placed on the exceptionally favoured situation of the U.S.A. rather than on the shrinking amount of space per head. At the present time, in crude density terms, the U.S.A. (conterminous states only), has less than 70 persons per square mile,[4] which is a far superior situation to that of most other large countries of the world. In China the crude density is about 200 per square mile, in Pakistan 325 and in the U.K. 590. If 'usable' or 'productive' land only is considered, the U.S.A. holds a very superior situation indeed, enjoying standards which set the country apart from most others. Even with a population of double the present size (1970, circa 203 million), the U.S.A. would still rank among the better endowed nations in respect of space *per capita*.

A rather more optimistic tone was therefore set in another departmental report of 1967. In *200 Million Americans* the Bureau of the Census briefly took up, and as swiftly dismissed, the fears expressed by Udall. 'Are there too many of us?...Will we be swamped by people, drowned in a population flood?' The view expressed here is that this kind of thinking was too highly coloured by the daily experiences of those living in large metropolitan areas. (Currently 70 per cent of the population lives on only two per cent of the land.) In fact, the 1967 report argues, there is much uncrowded space and people should 'beware of predictions of dire times ahead because of overpopulation'.[5] Between 1960 and 1970, according to the Census, almost half the counties of the nation *lost* population (Figure 1), while total

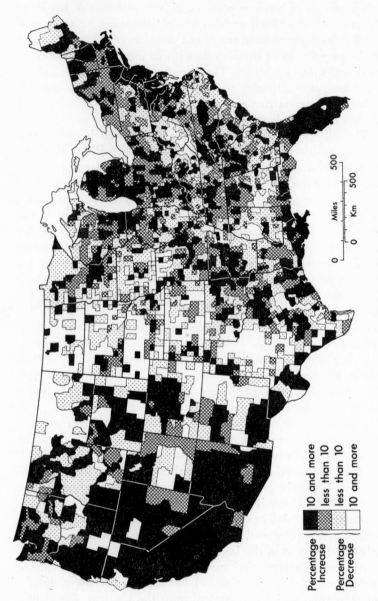

Fig. 1   Population change 1960–70

Percentage
Increase
10 and more
less than 10

Percentage
Decrease
less than 10
10 and more

500
500
Miles
0
Km
0

population rose by some 25 millions. A similar pattern was recorded for the previous inter-censal decade, and overall the amount of 'uncrowded space' may well be rising.

A realistic view of the situation would avoid both the extremes of alarm and complacency. One may legitimately reject the thesis of 'dire times ahead' while at the same time acknowledging that there are various difficulties—economic, social, environmental—which will accompany further growth and geographical shifts of population. The American problem, with the continuation of the growth rates of recent years, will not be for some time basically one of numbers *versus* resources overall. It is already, however, a problem of where, of how adequately and how efficiently, at what consumption standards and what cost to environment, it will be possible to cater for the increasing numbers. Population growth between 1960 and 1970 represents, actually or prospectively, a great volume of new demands for homes, schools, jobs, facilities for health and leisure and so on. The impact on resources and environment is bound to be considerable and raises urgent questions on the adequacy of preparations for dealing with further inevitable growth.

Moreover, the expanding numbers of Americans provide only one element in a complex man-environment situation. Other major elements include the levels of urbanization and of industrialization, locational issues, new technology and rising productivity. The last two combine to create what is probably the most potent element of all in the American man-environment context, the growing level of output and consumption *per capita*. Measured in constant 1958 dollars, *per capita* disposable personal income in 1950 was $1,600, rising to almost $2,600 by 1970. In other words, the real purchasing power of every American rose by above 60 per cent over these twenty years—and by 1970 there were 52 million more Americans commanding this extra spending power. *Per capita* purchases of goods rose from under $1,000 in 1950 to $1,400 in 1970 (constant 1958 dollars), representing a great increase in the demands upon resources of all kinds, with an inevitable environmental impact in the making. Several illustrations of the process of rapid scientific and technological advance which has helped to achieve astonishing levels of productivity will indicate the environmental implications.

In the agricultural sector in 1900 each farm worker produced sufficient to supply seven persons with their farm product needs. By 1940, modest productivity gains had raised the number supplied to 10·7, and swifter advance thereafter raised the total to 26 persons by 1960. In the 'sixties, however, the achievements of former decades were made to look commonplace for, by 1970, each American farm worker produced sufficient to support 47 persons. The environmental consequences (using the term in its broadest sense) have been enormous. With a greatly reduced demand for labour (Table 1), the U.S. farm population fell from 30 millions in 1940 (23 per cent of U.S. population) to 9·7 millions in 1970 (under 5 per cent of the population). The social and economic repercussions of this in farming areas need no elaboration, while the migration of millions from farm to city has often had effects little short of disastrous both for the cities and for the migrants themselves.

Table 1   Quantities of Selected Farm Inputs, 1950–70 (1950=100)

	Labour	Farm Real Estate	Mechanical Power & Machinery	Fertilizer & Liming Materials	All Others
1950	100	100	100	100	100
1960	67	98	115	169	129
1970	46	107	130	353	170

Source: U.S. Department of Agriculture, *Agricultural Handbook No. 423, Handbook of Agricultural Charts*, 1971.

Furthermore, the prodigious achievements in productivity were accomplished by greatly increasing the non-labour inputs to the land (Table 1). Thus the input of mechanical power and machinery in 1970 was 30 per cent higher than the already very high level of only twenty years earlier, a development which, with the changes required in farm method and organization, has considerable environmental implications. Of more immediate concern, has been the rapid increase in the application of chemical fertilizers and chemical methods of disease and pest control. In the single decade 1960–1970 fertilizer input rose by more than 200 per cent and in some major farming areas streams run heavily laden with nitrates and other chemicals. Chemical pesticides, the consumption of which approximately doubled during the 1960s, have proved a source of even greater anxiety because of suspected adverse secular effects on plant and animal life, farm workers and, possibly, consumers of food. The chain of consequences here could be most far-reaching, but it is still only partially understood and remains a matter of some controversy.[6] Overall, the productivity achievements in agriculture have been great and in many respects beneficial. As a result, Americans have been able to spend proportionately less of their incomes on farm produce, while others have benefitted from American scientific and technical expertise. But the environmental side effects have called for greater attention.

In the field of mineral production, technical advance has also greatly raised productivity and provided an important foundation stone for the rising affluence of U.S. society, but again with grave consequences. In 1950, for example, bituminous coal production amounted to 516 million tons, and in 1969 to 561 million (Table 2). Over the period average output per man-day

Table 2 Bituminous Coal Mining, Selected Data 1950–69

	Output (mill. tons)	Av. Number of Men Employed (thousands)	Output per Man-day (tons)	% Mined by Stripping
1950	516	416	6·8	24
1960	416	169	12·8	30
1969	561	129[1]	19·4[1]	35

[1] 1968 data.
Source: *Statistical Abstract of the United States*, 1971, Table 1049.

rose almost 300 per cent, and the labour force contracted by 70 per cent. Large areas are thus afflicted with coal-mining communities which have no *raison d'etre*, while the achievement of such high levels of output per man has had notable physical consequences. The advances in deep mines have been considerable, but have been far outweighed by the surface operations, which now account for above one-third of all coal mined. Average output per man-day in strip mines reaches 34 tons, and overburdens of 100 feet or more are mechanically removed to gain access to the seams.

The environmental consequences are profound. In flat or gently undulating country, square miles of land are turned over and abandoned in hill and dale formation, disrupting surface and sub-surface drainage patterns, exposing bed rock for subsequent leaching, and rendering the land both economically useless and devoid of aesthetic merit. In hill and mountain country the technique has the dignified label of 'contour mining', as seams are followed round the hillside, which is cut into as far as prudence and economics permit. Mile upon mile of contour bench and associated cliff face ('highwall') are left to isolate large hill-top areas, while huge volumes of waste are thrown down the hillside. Vegetation is thus destroyed, valuable top soil buried, land slides block streams and roads, waters become heavily polluted with acids and sediment, and fish and wildlife for miles around are poisoned. In Appalachia alone a 1966 Report estimated a total of 800,000 acres 'disturbed' by surface workings; 20,000 miles of highwall (often circumscribing and rendering inaccessible entire mountain top areas); and almost 11,000 miles of streams seriously affected by acid mine pollution.[7]

Land surface and sub-surface despoliation in mineral working is not confined to coal. The vast American appetite for minerals for energy, manufacturing, construction and agricultural use wreaks havoc over wide areas—and is not restricted in its effects to the U.S.A., since American demands call forth supply from many parts of the world. Coal is the largest single cause of what is officially called 'land disturbance', accounting for about 40 per cent of all such disturbed land in 1965.[8] Close behind, with some 37 per cent of the total, comes the working of sand, gravel, stone and clay. The remainder is shared principally by gold, phosphate rock and iron ore workings. The broad geographical patterns of this disturbance are shown in Figure 2. The damage done by surface coal working is massively concentrated in eastern, especially north-eastern, areas and here the landscape consequences described above are a major affliction. It is clear from Figure 2 that no state is entirely free of environmental difficulties of these kinds. The working of sands, gravels and

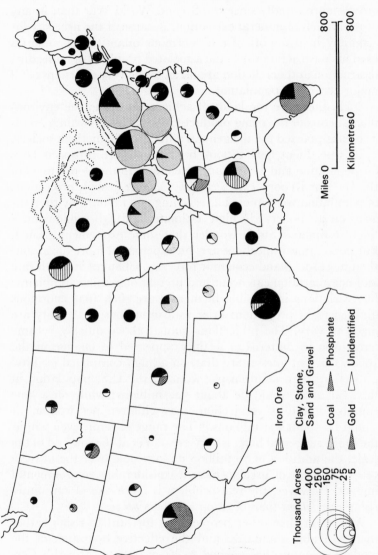

Fig. 2   Land disturbed by mining 1965

stones is a virtually ubiquitous activity in which output has grown more rapidly since the Second World War than in any other branch of mineral extraction. Because of the necessity for economy in transport of the enormous quantities of minerals used in construction work, the associated problems of noise, dirt, destruction and dereliction are taken to the very doorsteps of all major centres of population.

Such, then, are the kinds of achievements and the environmental costs of the rising productivity that is supporting, and is in turn supported by, the general affluence of American society.

It is pertinent next to look briefly at future prospects both for population and for consumer demand. In the Department of the Interior Report *The Population Challenge* (1966), projections of population were quoted as reaching 400 millions by the year 2010 on a 'moderately high' estimate, and 438 millions on a 'high' estimate.[9] Small wonder that some excitement ensued. But population projections are, on experience, more hazardous than most social and economic forecasts. Changes in death rate and net migration can occur and present only minor problems; but variations in possible birth rate are such that enormous differences in projections arise as assumptions on fertility are made to vary. The projections quoted above quickly became unrealistic as patterns of fertility appeared to change in the 1960s. Current guesses are that, on a 'high completed fertility' projection (3·10 children per woman) the U.S. population in the year 2000 would be about 322 millions; while on a 'low completed fertility' projection (2·11 children per woman) it would be about 271 millions.[10] The range of error, even within the limits suggested here, is very great. Yet at the low end of the scale, the addition of 70 million or so people over the next 30 years is bound of itself to have a considerable environmental impact. However, the real problem is more one of constantly rising *expectations* than of rising numbers *per se*.

Americans, like other people, look forward to rising rather than to stable standards, and the effective operation of the modern economy has seemed to be geared to 'growth'. Certainly it appears that economic expansion is necessary if urgent domestic problems are to be resolved. In 1969, for example, 24 million Americans were living below the 'poverty thresholds' (which ranged from \$1,797 for an unrelated individual to \$3,745 for a non-farm family of four)[11] and the economic difficulties of the early 1970s have resulted in a sharp increase in the numbers of unemployed. The current strategies against such domestic conditions seek to raise demand yet further. The alternative is that the better off accept a lowering of their own current standards, but discontent with current income levels is pervasive. R. K. Merton writes 'in the American Dream there is no final stopping point...At each income level...Americans want just about twenty-five per cent more'. Society's disciplines and pressures join 'to retain intact a goal that remains elusively beyond reach'.[12] Such a situation has explosive implications, and such inner drives re-inforce the search for new technologies and ever higher levels of production so that real purchasing power can be increased. Figure 3 relates the expansion of personal consumption expenditures to the growth of population over the period 1930–70, with the monetary values held at their 1958 dollar equivalent. The growth of expenditures outreaches that of population over every period, but recently at a very much more rapid rate. The projection of population growth over the period 1970–2000 reaches a figure of 280 millions, about 128 per cent above the base level of 1930. The line for consumption expenditures moves off the graph, but if maintained at the 1960–70 rate would reach about 260 per cent of the base level by the year 2000. This is not out of line with business predictions that median family income (on a constant dollar basis) will more than double between 1970 and 2000, while much more of the total available will represent discretionary income. The purchasing power of such a population would be alarming if directed, as at present, largely to material possessions. According to Herman P. Miller, the Chief of the U.S. Census Bureau, it is the pollutant power of family income more than population growth which will determine the state of the environment at the end of this century.

Another very important issue at the meeting ground of population and environment, and one compounded of population growth and rising real incomes, lies in land use, especially for urban-industrial purposes. Population growth in recent decades has been generally confined to urban areas. Currently

about three-quarters of the U.S. population lives in urban areas
or their immediate extensions, the number having expanded by
more than 50 millions since 1950. Clearly, the consequences for
land use are likely to have been severe. Unfortunately, American

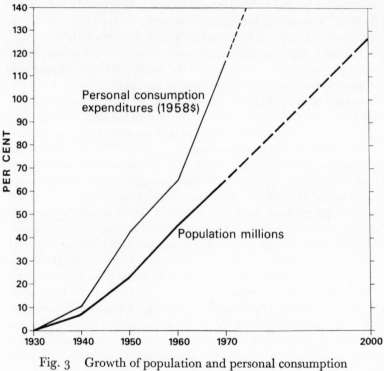

Fig. 3   Growth of population and personal consumption
expenditures

data on land use (and especially urban land use) are meagre and
unreliable,[13] while attitudes are often ill-informed and coloured
by nineteenth-century myths of endless resources. It is not
known with any accuracy how much land has been taken by
urban uses. An estimate by Resources for the Future Inc. for
1960 puts the area of land in 'cities' of 2,500 population and
above at 21 million acres.[14] A Department of Agriculture
estimate for 1964 placed 29 million acres in the urban use

category.[15] But however great the differences between estimates, under any reasonable definition of 'urban use', the total
is unlikely to exceed 2 per cent of the nation's total area. For
this reason it is sometimes difficult for Americans to see the true
nature of the land problems created by urban expansion. There
are two elements with important environmental implications
which are worth developing here—and there are the consequences of one myth to be conjured with.

Firstly, the rate of expansion of land in urban use has for
long exceeded the rate of growth of population. Between 1900
and 1960, population grew by 130 per cent but, according to the
previously mentioned study of Resources for the Future, total
land in urban uses rose by about 350 per cent. C. A. Doxiadis
examined the record for the 155 'urbanized areas' that existed
in 1950 (then accounting for 72 per cent of the total urban
population of the United States) and concluded that in the
decade 1950–60 there was a 30 per cent increase in population
for an 80 per cent increase in 'urbanized area'.[16] Moreover,
the flood of low-density suburban development which these
figures reflect has not in general been planned so as to achieve
an ordered or economical use of other land in the vicinity.
Much development is of the 'pepperpot' or 'ribbon' variety
which not only raises the costs of adequate servicing, but
impedes the effective use of areas not built over. Much land is
rendered idle and very large areas are held virtually derelict by
speculators in the hope of further urban growth. M. Clawson
estimates that the area of such idle land in, or on the fringes of,
urban areas equals the used area. He maintains that this
represents a typical requirement for 20 to 50 years of growth and
that while some such area available for development is essential,
the amount involved here is far in excess of any reasonable
quantity.[17] The structure of local taxation, which bears heavily
on buildings, but lightly on land, helps to perpetuate the
condition.

In reviewing the continuing impulse to low-density suburban
development, the consequences of the 'Jeffersonian myth' must
inevitably be considered. Jefferson's image of the virtues of
agrarian life and the essential rottenness of the city gained wide
acceptance in nineteenth-century America and led to dislike

and distrust of city life. Thus, while having perforce to seek the economic advantages of city life, modern Americans have sought to introduce a rural style of living into the urban context. D. J. Elazar suggests that the big cities continued to grow while they still catered for this desire. But 'when the problems of population density and congestion seriously cut into the possibilities for maintaining an agrarian-influenced life style within them', big city decline began. At that stage (in most places beginning in the inter-war period) smaller settlements and scattered suburban development became paramount.[18] If this judgement is accurate, the 'agrarian myth' has remained a force to be reckoned with. Mayor Lindsay reviewing the problems of New York after his election to a second mayorial term in 1969, placed it firmly at 'the root of the city's ills'. '(It) is on the face of it absurd...Nevertheless it is historically true (and) my thirteen years in public office...and now as Mayor of the biggest city in America, have taught me all too well the fact that a strong anti-urban attitude runs consistently through the mainstream of American thinking.' This not only affects past attitudes and legislation, but remains a powerful element to this day.[19] It is not, however, the entire explanation of the suburban phenomena in recent times. Other elements such as the Federal government's post-war mortgage assistance plans and the 1954 Supreme Court decision in favour of school desegregation have also played a part. But one danger of the agrarian myth, insofar as it remains a real force, lies in an apparent belief in the fallacy that rural life is rural because densities are low. Consequently, as millions upon millions have sought 'rurality' in low density sprawl, they have inevitably destroyed much of what they sought. In addition they have created a host of new problems for themselves, their children and the great cities from which they moved—problems at once economic, social and environmental. In terms of land area only, however, according to Department of Agriculture estimates, urban land uses have been taking about one million acres of new land annually. Given the size of the country this may still seem negligible. An additional area the size of England could thus be covered by the year 2000, yet total urban acreage only amounts to a little over 3 per cent of the whole. At this point the second major environmental element becomes significant, notably the geographical location of the new urbanized territory.

Currently there are four major urbanized zones which account for about half the urban area of the nation. These are the North East coast, from Massachusetts to Maryland; the 'Lakes' area, roughly outlined by a line joining Cleveland, Detroit, Chicago, Cincinnatti and Pittsburgh; southern coastal California; southern and eastern Florida. Without doubt these peripheral areas have been, and will continue to be, those where suburbs and satellites expand most rapidly. Figure 1 shows that these areas remained vigorous centres of population growth between 1960 and 1970, experiencing high rates of growth despite their very large base date populations. This map, drawn on a county area base, identifies high growth rates elsewhere, but these have normally been on relatively small 1960 base populations. For example, Arizona and Nevada appear to have had very high growth rates over much of their area,[20] but together they raised their populations in the decade by only 670,000. Similarly in Oklahoma, where strong growth is shown to have occurred in northern and western counties, the total expansion amounted to only 232,000. By contrast, the peripheral areas identified above experienced very large absolute increases. If data for whole states are employed, during the decade 1960–1970, Illinois, Indiana, Ohio, and Michigan added 3·6 million to their populations; the coastal states from Massachusetts to Virginia, 5·8 millions; Florida 1·8 millions and California 4·2 millions. Altogether these peripheral growth zones accounted for nearly two-thirds of the total increase in U.S. population in the period. Writing in 1967, J. P. Pickard estimated (correctly) that by 1970 these areas would contain about half the nation's population, on but 5·4 per cent of the land area of conterminous U.S.A., at an average density of 630 per square mile. These same areas, he argued, would command 80 per cent or more of population growth to the end of the century, and would then house 60 per cent or so of U.S. population (which he assumed to be 312 million) at an average density of 822 per square mile. The density would rise despite an increase by some 55 per cent in the area of these great urban zones.[21] Such a concentration of population has its favourable aspects, as Pickard emphasizes.

Very large areas will be preserved as open country with only occasional urban centres, and outside the main zones the average density of population might even decline. This 'reserve of space' would be of immense value. Even in the great urban zones, moreover, there would continue to be considerable open space. The regions concerned would be dominated in every way by a continuous succession of metropolises and smaller cities, but they would not be completely built over. What is made of such 'internal' open space would be, of course, a matter of effective planning and ordered development—no easy things to achieve in the United States. However, as Pickard also stresses, the massive concentration of people into 'systems of metropolitan areas and high density urban zones poses many problems in terms of water supply, waste disposal, environmental pollution and other factors', and these are matters which today excite rising concern.

We have, therefore, in this theme of population growth and environment a series of considerable problems. What of the response to these changing conditions, to these dilemmas of growth and rising affluence? Manifestly, the feeling that population growth might be a burden rather than a boon has advanced rapidly since the mid 1960s. At the same time there has been an increasing awareness of various issues relating to environment. This is not to suggest a total lack of concern in previous decades, but those who involved themselves in environmental issues at that time were often dismissed by the general public as 'scare-mongers'. Nonetheless they had occasional impact and accordingly we find the current surge of concern over problems of resource and environment described as 'The Third Wave', a title adopted by the Department of the Interior for its third Conservation Yearbook, 1967. The 'first wave' is regarded as having coincided with the Presidency of Theodore Roosevelt and the administrative ascendancy of Gifford Pinchot (1905–10). It receded swiftly when Roosevelt left office, having had its greatest impact on the forest lands. The 'second wave' embraced the F. D. Roosevelt New Deal period, although enthusiasm for conservation was perhaps somewhat subservient to the use of conservation needs to provide jobs and stimulate economic recovery. Both 'waves' thus owed much to the character of individuals, while aims were limited and not necessarily confined to conservation issues. In essence, too, neither 'wave' had a broad base of public enthusiasm and support.

The 'third wave', beginning in the early 1960s and since gathering momentum, has been broader in concept and based much more on popular support. The earlier philosophy of conservation of resources for their wisest use has broadened to embrace all aspects of human and physical environment (preservation, restoration, improvement etc.) as well as the more nebulous concept of quality of environment. President Kennedy succeeded in attracting public interest in these matters, but he achieved little on the statute book. His appointment in 1961 of Stewart Udall as Secretary of the Interior, however, placed in a pivotal position a man deeply concerned with conservation and environment, and whose public statements during a long tenure serving both Kennedy and Johnson attracted wide attention by their forthright (some would say alarmist) quality. The Conservation Yearbooks of 1965–8 reflected his concern, and the rising element of public interest and support was, in turn, reflected in a considerable volume of relevant legislation during the Johnson presidency.

Rising affluence has not only created numerous and obvious manifestations of drastic environmental deterioration, but also provided some of the wherewithal from which the problems could be tackled. In the boom conditions of the 1960s more people came to feel that such a prosperous nation ought to be able to direct resources towards combatting environmental pollution and destruction, without drastic repercussions on the material standards of individuals. Only the will to do so needed to be strengthened. At the same time, the conservation movement was aided by the creation of a national constituency in these matters. Modern communications have broadened horizons so that major environmental problems and conflicts are no longer faced and fought at purely local level. Oil leakages at Santa Barbara, the eutrophication of Lake Erie, temperature inversion and killer smog in New York or Los Angeles have become recognized as national problems. In local conflicts, the internal pressures which because of jobs and income often work in favour of exploitation (even at considerable environmental cost) are now opposed by outside pressures which, with no personal financial interest, normally favour conservation.

One example of the operation of these pressures was seen in the late 1960s when a plan to construct a large new airport close to the Everglades National Park, a plan strongly supported by powerful interests in Florida itself, was prohibited by the Federal authority. Currently the trans-Alaska pipeline controversy demonstrates the power of the environmental lobby. With oil production in the conterminous states unable to cope with current—let alone future—demands, an alarming increase in the dependence upon foreign supplies is in prospect.[22] This could be reduced in size and urgency if the extensive oil fields on the Alaskan North Slope, which contain a quarter of all the proved recoverable reserves in the U.S.A., were able to begin production. But operations await the provision of transport facilities. While the great oil companies favour the construction of a pipeline across Alaska from Prudhoe Bay to Valdez, the opposition of environmental interests has already delayed the project for several years. Legislation outlined below requires that a full analysis of the environmental impact of such major projects be published before official approval is given. The environmental lobby has pressed this requirement to the full, rejecting as inadequate a study made for the Department of the Interior in 1970, and forcing the Secretary of the Interior to conduct a much fuller investigation. The investigation was completed and published in March 1972. It is a monumental nine-volume work[23] over which long and complicated legal battles will have to be fought before construction permits are finally issued and work on the line begins. Whatever the outcome the wide awareness of environmental problems and the effective involvement in conflict of environmentalists are new American experiences and a national constituency in favour of conservation is being formed. Given this new situation, no large or national interest group can openly oppose pollution control and environmental improvement legislation. Indeed, much advertisement aims to show how environmentally conscious big business is.

Such new expectations place some groups and interests in positions of great difficulty. For example, many large manufacturing industries (oil, chemicals, iron and steel, cement, motor vehicles etc.) either in their processes or in their products gravely damage the environment. Pollution control is costly and of itself adds nothing to the value of the product. Moreover, huge capital sums are invested in existing production facilities which may not lend themselves to the adoption of new pollution control methods and devices. One estimate of the total spending required in major fields of pollution control over the period 1970–5 puts the total at about $105,000 millions; 23 per cent of which would be for air pollution control, 36 per cent for water and 41 per cent for solid waste.[24] Well over half this enormous sum would accrue to the activities of private enterprise.

Table 3 Pollution Control Expenditures, Selected Industries 1970 ($ mill.)

	Pollution Control Air	Water	Pollution R and D	Capital Expenditure on Pollution Control as % of Total Cap. Exp.
Petroleum Products	152	185	34	6·0%
Iron and Steel	86	120	2	10·3
Chemicals	79	90	53	4·9
Paper	59	94	8	9·3
Machinery	82	39	178	3·5
Non-ferrous metals	80	20	10	8·1
Instruments	18	7	32	3·6
Aerospace	9	6	181	2·8
All Manufacturing	847	872	661	5·4
Non-manufacturing business[1]	496	287	n.a.	3·1[2]

[1] Mining, transport, utilities etc.   [2] including manufacturing.
Source: *Council on Environmental Quality, Second Annual Report*, 1971, pp. 82, 85, 126.

Table 3 shows that manufacturing industry spent more than $1,700 millions on air and water pollution control alone in 1970, and a further $660 millions on Research and Development programmes in pollution control. Outside of manufacturing, activities such as mining, transportation, gas and electric utilities spent $780 millions on air and water pollution control. Such expenditures were being made before the pressures for improvement had become heavy, and outlays were expected to rise markedly in 1971 (by 46 per cent according to one estimate),[25] and thereafter even further. Table 3 shows that the burden varies greatly among the major industry groups while, even at the relatively low levels of spending in 1970, in several cases capital expenditure on pollution control rose to more than 5 per cent of total capital expenditures. Within each industrial group the ability to meet new outgoings of these types will vary greatly from firm to firm and plant to plant. Most will be able to absorb, or pass on, the costs. But for many, despite the availability of various schemes of federal aid, these added requirements will prove ruinous so that the entire economic base of some communities will be undermined.

On the other hand it must be recognized that there is arising a group of new business activities and enterprises which will *make* money out of pollution control and environmental improvement work, and that their activities will create new jobs and help enlarge the Gross National Product. It will be noted, for example, in Table 3 that the aerospace industry (itself a minor pollutor) accounted for well over a quarter of all expenditure by manufacturing industry on pollution control research and development in 1970. The machinery group is similarly very heavily committed in this field. This is no accidental process, but a deliberate attempt by firms in these branches of activity to broaden the base of their operations and to capitalize on the new growth industry of pollution control. Vast sums will be spent and a large new interest group will emerge whose own business *raison d'être* will depend upon the continued and advancing public interest in pollution control. The weight of this new lobby as a business sector, allied with that of the crusading element whose growing influence has already been emphasized, will fall on the environmentalist side to create a powerful voice in Washington.

Given the size and nature of environmental problems, and the fact that the consequences of environmental damage can rarely be geographically confined but flow freely across man-made boundaries, it was inevitable that the Federal government should become increasingly involved. Individual efforts by state and local authorities can be nullified by indifference in neighbouring areas. Furthermore, in a competitive situation it is impracticable for areas in isolation to enforce on their own populations and business enterprises environmental quality standards which are reasonable, yet costly. The clear requirement was for Federal intervention to set standards, enforce compliance, and assist in various other ways. Thus, given the developing attitudes of the 1960s it is not surprising that a considerable volume of 'environmental' legislation was achieved during the Johnson Presidency. The two major fields of endeavour were in air and water pollution control, a fact which is reflected in their receipt of some 88 per cent of Federal fund obligations for pollution control and abatement in 1971.

Limited attempts at water pollution control were made before the Second World War, and again in 1948, but the first permanent legislation came as recently as 1956 in the Federal Water Pollution Control Act. This has been subsequently amended, and supplemented by the Clean Water Restoration Act of 1966 and the Water Quality Improvement Act of 1970. In essence this legislation permits the Federal Government to participate with, and assist, the State Governments in programmes of water quality control, and to supervise and approve the plans and standards set in each state. However, the enforcement mechanisms provided under the Acts have been described as 'limited and cumbersome',[26] and late in 1970 the President announced that recourse was to be made to the Refuse Act of 1899, which outlaws discharges and deposits into all navigable waters, unless a permit has been obtained. The use of this legislation enables swifter action to be taken against water pollutors, and allows a precise guide to be given on the nature and extent of permissible pollution. Even more rigorous measures are currently being debated in Congress. Together with the various enforcement mechanisms and backed by the persuasive power of the Federal purse, they should be adequate to achieve some improvement in water quality.

Air pollution, for its part, was not recognized in any way as a problem until the post-war years, and it was not until 1963 that the Clean Air Act (the first permanent piece of legislation of this character) was passed. This was superseded by a more comprehensive Air Quality Act in 1967, which was itself fundamentally amended in 1970. The Federal Government now has power to establish air quality standards, both nationally and for individual sources of air pollution (e.g. motor vehicles). The duties and responsibilities of state governments in this respect are set down and, when the Federal agency has finally established the standards, the Government will possess enforcement powers, as well as the right to provide financial and technical support.

Additional legislative measures have embraced other aspects of the environmental problem, such as pesticides, noise, radiation, solid waste disposal, the provision of recreation areas. The one notable omission has been in the field of land use. Yet a national land use policy, involving the mandatory construction and approval of land use plans in local areas, appears essential both to reduce the profligate use of land and to achieve a better balance and arrangement of uses (housing, industry, transport, mineral working etc.) in developed areas. Various official powers over land use already exist, as in the 'police power' (permitting zoning and subdivision control), in the right of eminent domain, in tax policy and in government programmes involving the use of land. But such powers are not applied consistently and their existence and piecemeal application does not add up to a land use policy. Proposals have been laid before Congress and their urgency stressed by the President,[27] but nothing has yet been achieved.

Moreover certain weaknesses soon became evident in the various threads of government activity which had been spun. It can be argued that the legislators were too busy in the 1960s producing too many pieces of popular environmental legislation with far too little commitment of real resources. As in other programmes of the Great Society era (e.g. distressed areas, economic opportunity, urban development) the financial commitment did not measure up to the scale of the problems and the stated objectives of the legislation.[28] Again, the numerous enactments resulted in a considerable dispersal of effort, duplication of programmes and fragmentation of control. The practice of allocating programmes to existing departments of the

administration produced problems in its own right since, in the
nature of things, a number of different departments could claim
a legitimate interest in any given programme. Interdepartmental
friction was bound to result. Moreover, the many separate
enactments gave to a variety of agencies powers which over-
lapped or even conflicted. J. C. Davies identifies 20 Federal
agencies involved in air pollution control and more than twice
as many concerned with water resources.[29] Again, agencies which
were regarded as being in general charge of a specific field of
pollution control have sometimes been those most concerned to
advance the interests of the polluting activity itself. Thus the
Atomic Energy Authority had the right to set and enforce
radiation exposure standards at atomic energy plants, and
throughout the 1960s consistently played down the hazards
involved. Attempts are now being made to tackle some of these
difficulties and to express in a more formal way the national
government's responsibility for and involvement in these
matters.

The 1969 National Environmental Policy Act provides a
major landmark, perhaps being comparable in its field to the
famous 1946 Employment Act. The latter had declared it to be
the 'continuing policy and responsibility of the Federal Government to use all practicable means…to promote maximum employment, production and purchasing power'. This was what
Johnson later called 'the essential and revolutionary declaration'
upon which subsequent developments in Federal policy and
practice could be based. In the 1969 Act, Congress recognized
the profound impact of man's activity on the environment, and
the critical importance for the overall welfare and development
of man, of restoring and maintaining environmental quality. It
was declared as the 'continuing policy of the Federal Government…to use all practicable means…to create and maintain
conditions under which man and nature can exist in productive
harmony…' In a sense, the objectives of these two major enactments conflict, since expansion of output and purchasing power
have adverse environmental implications. Clearly, however, what
is implied in the 1969 Act is the need to pursue economic growth
with the goals of a good environment also in mind. Under
the Act the *right* of every person to a healthy environment

and the *duty* of every person to contribute to its preservation and enhancement are stressed. Federal agencies are charged to examine the environmental impact of all their programmes, publish their findings and issue specific directives on the prevention of damage. A Council on Environmental Quality (similar to the 1946 Council of Economic Advisors) was established to report to and advise the President on relevant issues, and to present an Annual Report to Congress.[30]

An important reorganization of environmental affairs followed in December 1970 with the establishment of the Environmental Protection Agency. This independent agency (reporting directly to the President) was to take over responsibility for clean air, the water programmes, solid waste management, pesticides control, radiation standards and research on ecological systems. It thus pulled together a variety of work and enforcement activities previously scattered across several agencies, while future environmental programmes will automatically come under its control. Also in 1970, President Nixon set up a Commission on Population Growth and the American Future to report on the broad range of population problems.[31]

To conclude, there is no doubt that the U.S.A. is responding to the environmental problems set by population growth and rising consumption. The question is whether the response so far, and the real commitment of resources, is sufficient. In closing their book on *Congress and the Environment*, for example, R. A. Cooley and G. Wandesforde-Smith write 'The major conclusion is inescapable and bitterly disappointing...Congress has failed to do more than make halting progress through a series of incremental adjustments. Every chapter in this book points to serious flaws in the response of Congress to the basic moral and political issues posed by (this) crisis.'[32] But even under the most favourable circumstances, the 'response of Congress' can only provide a partial answer to the problems. Pollution and environmental destruction cannot easily be legislated out of existence when the chief cause is the very high levels of production and consumption which politicians must promise to sustain, and advance, if they are to obtain power. The fact is that, to support the present level of demand for goods, an extraordinary burden is being placed upon the environment. The problem is heightened because the impact of America's gargantuan appetite is global since it draws heavily upon resources from elsewhere. It is estimated for example that, with some 6 per cent of the world's population, the U.S.A. currently consumes one-third or more of the total world output of minerals. Solutions depend in a large measure upon a more realistic appraisal of human needs and a more conservative approach by individuals to the use of resources—changes in attitudes of mind and philosophy that will not easily be made.

In closing, then, we return to an earlier theme. It is not only considerations of population size that have thrust matters of environment into the forefront of concern, but also human attitudes, technology, productivity and high *per capita* demands. There is a relatively straightforward, although not easy, answer to population growth in reducing the size of families. Current developments in the U.S.A. appear to indicate that natural increase will continue at present low levels so that increments to population will be less a matter of large families than of larger numbers of existing people reaching childbearing age. The secular trend may already be towards a very slowly growing population. There is no straightforward answer to the variety of problems created by production technology and the appetite to consume. Some have suggested the desirability of retreating to a lower level of technology. But to achieve any notable result, the retreat would need to go far since, relative to population size, man was already exerting a very deleterious impact a century or so ago. Thus a retreat to the technological level of 1870 would in the first instance reduce the majority of Americans to well below the current poverty threshold. In the second instance, it would not protect the environment since pollution and destruction were already at high *per capita* levels in 1870.[33] Rather, perhaps, hope lies in the opposite direction—that the appetite to consume might be increasingly directed towards consumption patterns which place less strain on environment, that consumers may be persuaded to spend more of their income *on* environment, and that continuing technological advance will provide the means for coping with the problems that it has itself created.

NOTES

1. J. Russell Smith and M. Ogden Phillips, *North America*, 2nd edn., Harcourt Brace and Co., 1940, p. 1.

2. Ibid.

3. Department of the Interior, *Conservation Yearbook* No. 2, 1966, p. 3.

4. 58 per square mile if Alaska and Hawaii are included.

5. Department of Commerce, Bureau of the Census, 200 *Million Americans*, 1967, p. 3.

6. Dr N. Borlaug, a 1970 Nobel Peace Prize winner, addressing the U.N. F.A.O. Conference in Rome in November 1971, mounted a strong defence of the use of chemical pesticides, and attacked their critics as ill-informed and hysterical. A reply by M. Allaby is published in *The Ecologist*, **2**, 4, April 1972.

7. Department of the Interior, *Study of Strip and Surface Mining in Appalachia, An Interim Report*, 1966.

8. Department of the Interior, *Surface Mining and Our Environment*, 1967.

9. *Conservation Yearbook* No. 2, op. cit., pp. 9–10.

10. Bureau of the Census, *Population Estimates and Projections*, Series P-25, 470, November 1971.

11. Bureau of the Census, Current Population Reports, *Consumer Income Series*, P-60, 76, December 1970.

12. R. K. Merton, *Social Theory and Social Structure*, The Free Press, New York, 1968, pp. 190–1.

13. See M. Clawson and C. L. Stewart, *Land Use Information*, Johns Hopkins Press for Resources for the Future, 1965.

14. M. Clawson, R. B. Held, C. H. Stoddard, *Land for the Future*, Johns Hopkins Press for Resources for the Future, 1960.

15. Dept. of Agriculture, *Major Uses of Land and Water for 1964, Agricultural Economic Report*, 149, 1968.

16. C. A. Doxiadis, *Emergence and Growth of an Urban Region*, 1, *Analysis*, Detroit Edison Company, 1966, pp. 57–9.

17. M. Clawson, *Land for Americans*, Rand McNally, New York, 1965, p. 16.

18. D. J. Elazar in R. A. Goldwin (ed.), *A Nation of Cities*, 1967, pp. 104–5.

19. J. Lindsay, *The City*, The Bodley Head, 1970, p. 50 and Chapter II.

20. The size of counties in these western states is often much greater than in the east so that population growth in relatively small pockets appears on the map to cover large areas.

21. J. P. Pickard, 'Future Growth of Major U.S. Urban Regions', *Urban Land*, **26**, 2, February 1967.

22. Estimates at the degree of dependence on foreign oil supplies by 1980 vary from about 50 per cent at the minimum to about 80 per cent.

23. Department of the Interior, *Environmental Impact Statement*, 1972.

24. Council on Environmental Quality, *Second Annual Report*, 1971, pp. 110 ff.

25. Ibid., p. 83.

26. Ibid., p. 12.

27. See, for example, President Nixon's Message on the Environment, sent to Congress, 8 February 1971, Part III.

28. In the 1970s, Federal funding for pollution control and abatement has risen. Budget authorization rose from $1,432 millions in 1970 to an estimated level of

$1,828 million in 1971 and $3,127 millions in 1972. See Council on Environmental Quality, *Second Annual Report*, 1971, Table O.2, p. 338.

29. J. C. Davies, *The Politics of Pollution*, Pegasus, 1970, p. 99.

30. The Council was set up in January 1970, and transmitted its First Annual Report to Congress in August 1970.

31. A brief Interim Report, *Population Growth and America's Future*, was published in March 1971.

32. R. A. Cooley and G. Wandesford-Smith, *Congress and the Environment*, University of Washington Press, 1970, p. 227.

33. See *The Environment. A National Mission for the Seventies*, The Editors of *Fortune*, 1971, Chapter XIII.

An earlier version of this essay appeared in the *Journal of American Studies* 6, No 1, April 1972 and is reproduced with the kind permission of the editor.

The Vanishing Cornucopia?

Some reflections upon the fuel and mineral wealth of the United States

GERALD MANNERS

In providing an abundance of readily exploitable natural resources for the first generations of European settlers in the North American continent, the unquestioned generosity of nature powerfully conditioned the attitudes of both them and many subsequent commentators on the American scene. The impression of virtually inexhaustible natural riches—of unlimited land, unending forest, abundant wildlife and unparalleled mineral wealth—was reiterated throughout the length and the breadth of the United States and took firm root in its national folklore. Isolated voices forcefully protested against some of the more wasteful reactions to this abundance. An influential conservation movement had appeared by the turn of the present century, and men began to challenge the not inconsiderable misuse of land, the wasteful clearance of the forest and the careless exploitation of mineral deposits. It was in fact during the administration of President Theodore Roosevelt that extensive areas of forest lands were deliberately set aside or withdrawn from use to become what are now the much treasured National Forests.

The image of an exceedingly abundant American resource base nevertheless survived well into the present century and it is an image that is not without some currency today. After all, the United States remains universally acknowledged as one of the finest sources of coking coal in the world. By 1970 annual coal and anthracite production had once again topped 600 million tons and of this some 70 million tons were exported. The country's coal reserves remain enormous, exceeding 1,500 billion tons.[1] Whilst the traditional 'direct shipping' iron ores of

the Upper Lakes region of the Mid West have been seriously
depleted and have become progressively less attractive to the
iron and steel industry, the 1960s saw huge deposits of taconite
ore opened for the first time in Michigan and Minnesota following the development of beneficiating and pelletizing technologies which permit their economical use. Copper production
has registered a significant increase in Arizona in recent years;
the United States remains the world's largest producer, with
26 per cent of world output in 1970 and with a considerable
share of the world's known copper reserves. The United States
is also the world's largest producer of lead (572 thousand tons
in 1970), and has opened up new lead deposits in Missouri
during the last 20 years to add at least a further 25 million tons
of the metal to the nation's reserves. The wealth and the
expanding horizons of the American natural resource base are
not difficult to exemplify.

However, it is noteworthy that the historical position of the
United States as a net exporter of resource products shifted to
that of a net importer over 40 years ago. While the country has
retained an advantageous balance of trade in basic agricultural
commodities (such as wheat and cotton), since 1930 it has
come to rely to an increasing degree upon foreign sources of
crude oil, lumber and a widening range of metals such as copper,
lead, zinc and iron ore.[2] Soon after World War II the United
States became for the first time a net importer of oil and iron
ore, and by 1950 the country was dependent upon imports for
more than half of its supplies of bauxite, manganese, nickel
and tin. These developments were monitored cautiously by the
National Resources Planning Board between 1933 and 1943;
but it was only after 1946—following a sudden upsurge in the
world demands and market prices for many basic raw materials
—that the need was felt for a definitive inquiry into the position
of the country's natural resources. This was provided by the
President's Materials Policy Commission,[3] sometimes known as
the 'Paley Commission', which reported in 1952 and looked
ahead to 1975. The Commission's work was paralleled and
supplemented by a number of more specialized sectoral studies
made by various government departments and agencies; and
in 1963 a second major inquiry—looking forward to the year

2000—was completed by Resources for the Future, Inc.[4] The recurring conclusion of all these resource inquiries was that, although there was a possibility of occasional scarcities of particular resources, such as lumber, at particular places and at particular times, the overall resource position of the United States was highly favourable and gave no cause for alarm. Has such a conclusion stood the test of time?

In both relative and absolute terms the United States in recent years has found it increasingly desirable—if not absolutely necessary—to rely upon foreign sources for many of its key industrial raw materials and sources of energy. Despite mandatory quotas, for example, by 1971 crude and refined oil imports into the United States totalled 1·4 billion (U.S.) barrels, approximately 27 per cent of the domestic market.[5] It has been estimated that, under certain price assumptions, the removal of the mandatory oil import quotas could lead to foreign oil capturing at least 50 per cent of the American market and the displacement of between one-third and one-half of the domestic producers, especially the high cost stripper wells.[6] In 1970, iron ore imports into the United States stood at 45 million tons. With exports of ore to Canada and Japan totalling 5·5 million tons, these figures represent net imports equivalent to nearly one-third of the iron and steel industry's ore requirements on an actual tonnage basis—and a somewhat higher percentage in terms of the ore's iron content. Imports of natural gas from Canada, Mexico and Algeria—and prospectively from the U.S.S.R.—have begun to rise steeply. Bauxite, tin and sulphur imports have also quickened in recent years. Zinc and chromium have been added to the list of minerals for which the United States relies upon overseas sources for more than half of its requirements—and by 1985 it has been forecast that oil, iron ore, lead and tungsten could be added also (see Table 1). Such forecasts raise as many questions as they answer, but to these particular forecasts have been added the more general warnings of some environmentalists and system analysts that not only the United States, but indeed the whole world, is on the verge of a major resource crisis.[7]

How then should these conflicting impressions and interpretations of the current and prospective natural resource position of the United States be interpreted? Is there still available to the American economy an exceptionally bountiful resource base, which for some reason has been neglected in the sweeping and far from rigorous assumptions of the environmental Jeremiahs? Or is the Cornucopia nearly empty? The purpose of this essay is not to provide a quantitative revision or refutation of the

Table 1   United States: Dependence upon Imported Minerals and Fuels, Actual and Forecast (percentage imported, by weight)

	1950	1970	1985*	2000*
Aluminium	64	85	96	98
Chromium	...	100	100	100
Copper	31	x	34	56
Iron Ore	8	30	55	67
Lead	39	31	62	67
Manganese	88	95	100	100
Natural gas	—	4	30	50
Nickel	94	90	88	89
Oil	8	22	60	65
Phosphates	8	x	—	2
Potassium	14	42	47	61
Sulphur	2	—	28	52
Tin	77	62	100	100
Tungsten	...	50	87	97
Zinc	38	59	72	84

— nil   ... not known   x net exports   * forecasts
Sources: *New York Times*, 5 November 1972; Dept. of Commerce, *Statistical Abstract of the United States, 1972*, Government Printing Office Washington D.C., 1972.

various published forecasts on resource availabilities to the economy of the United States. To usurp the role of the Federal government's agencies,[8] of industry organizations,[9] or of such distinguished private research corporations as Resources for the Future Inc.,[10] would indeed be foolhardy. Instead, the essay seeks to provide a perspective on the interpretation of resource appraisals in general and then singles out one resource commodity—oil—for detailed discussion.

In any natural resource evaluation, and more particularly in the appraisal of non-renewable fuel and mineral resources, three elementary yet fundamental considerations must be borne in mind. The first is the need to separate short term events from medium and longer term trends and considerations. It is imperative that resource appraisals see beyond the immediate fluctuations and vagaries of the market, which are so richly informed with industry information and press comment, and seek to delineate the more permanent tendencies over a 10 to 15 year period. The contemporary 'energy crisis' in the United States affords a good illustration of this situation and of the difficulties involved.

An extraordinary coincidence of unique events bore upon the market for energy in the United States between 1970 and 1972.[11] First, there was a serious misjudgement of energy demands by the major producing and supplying industries in the market and mixed economies, including the United States (which accounts for more than one-third of global energy consumption). There was then an interruption to the flow of Middle Eastern oil resulting from the temporary closure of the Trans-Arabian pipeline in Syria, followed shortly afterwards by the embargo upon a significant share of Libyan exports. Partly as a result of the former, there was a global shortage of ocean tanker capacity and a hardening of freight rates. The Trans-Alaska pipeline was being delayed by the environmental lobby. That same lobby emerged with unexpected strength to insist upon the more widespread use of low-sulphur fuels in many cities in the United States. A shortage of railway waggons limited the ability of the coal industry to meet a sudden upsurge of demand. Major delays were being experienced in the commissioning of many nuclear power plants in America. And the reserve and supply position of the natural gas industry began to deteriorate after fifteen years of federally-imposed low field prices for that fuel. All these and other factors conspired to make energy relatively scarce in relation to American demands, for its market value to rise, and even for physical shortages to occur in some places.

Clearly, the effects of these circumstances could not be reversed overnight. However, by the time President Nixon transmitted a special message to Congress on the country's energy problems in June 1971—the first occasion when a Chief

Executive has addressed himself solely to the topic in such a public fashion, thereby keeping the 'energy crisis' a continuing national issue—changes had begun to take place. The Trans-Arabian pipeline had been re-opened, the Libyan situation was returning to normal, additional tanker capacity was becoming available and freight rates had begun to soften; special arrangements had been made to increase oil imports from Canada and other sources, and penalties had been imposed upon the misuse of railway waggons, with the result that more were becoming available to move coal. Energy supplies remained tight into 1973, and the outbreak of hostilities in the Middle East added a further dimension to the problem. In time, however, it is not unreasonable to expect that the delays in the commissioning of United States' nuclear power plants will be overcome; that low-sulphur fuels will become more readily available, or the interpretation of regulations concerning air pollution will be somewhat modified in some places; that, over a longer time-span, additional supplies of natural gas will be found within the United States under new exploration and production incentives, at the same time as imports increase from overseas. In time, too, the international market for oil will doubtless absorb the impact of the Tehran and Tripoli agreements, plus the subsequent tax and ownership positions negotiated between the OPEC countries and the international oil companies. Indeed, depending especially upon the actions and longevity of the producers' cartel, it is not impossible that a buyer's market could begin to re-emerge by the middle of the decade. As the supply stringency weakens, energy prices will undoubtedly rest at somewhat higher values in actual dollar terms than before the crisis. It is more difficult to judge whether, in a world of rapid inflation, shifting exchange rates and unpredictable political events, the longer term trend for real energy prices to fall has at last been permanently reversed.[12]

For the purpose of this essay, however, the point to be established is that the evaluation of energy and other mineral resources needs to be placed within a medium term context; forecasts must look beyond the immediate situation and contemporary crises and events.

A second fundamental consideration in resource appraisal

also concerns the interpretation to be placed upon the most readily available information, and in particular the size of published energy and mineral reserves. Resource based industries in the market and mixed economies are instinctively reluctant to prove the existence of exploitable reserves in excess of their medium term needs. Whilst the management and the owners of the industry's firms are naturally anxious to be assured of the availability of their raw material requirements over a 10 or 15 year period, and whilst they might legitimately be apprehensive about supplies over a longer period, there are few financial incentives for them to peer too far into the future. The explanation is simple and economic. Little advantage can be gained from making the huge investments required for detailed geological and economic surveys and evaluations of particular deposits if the reserves subsequently proved therein will not be required for 20 years or more. Exploration costs can represent one-third of the total expense involved in finding a mineral, developing the appropriate production facilities and then actually winning the resource. Since all future values in a market economy are subject to a discount at the going rate of interest, a mining company (which is generally investing its funds with an opportunity cost of capital exceeding 15 per cent, is writing off its equipment over a 5 to 8 year period) cannot possibly value highly a reserve which will not be needed for 20 years or more. Even at a low 5 per cent rate of interest, a $100 asset to be realized in 20 years is worth only $35 today; at 10 per cent, it is worth $15; and at 15 per cent, the present worth of the future $100 asset is a mere $6.

With future resources valued by mining interests in this way, it is inevitable that knowledge of the true size of a region's or a country's mineral wealth is much scantier than is generally realized—even in countries as advanced technologically as the United States. It is not surprising, too, that major new discoveries of mineral and fuel resources continue to be made, and that reserves are persistently revalued upwards. Hubbert noted in connection with oil reserves that 'For the last 40 years the curve of cumulative production has faithfully followed the curve of cumulative discoveries with a time delay rarely outside the range of 10–12 years'.[13] There is nothing surprising in this

observation; it simply reflects the normal practice of any producer to adjust his level of inventory to the magnitude of his
production or sales.[14] Against this background, it is equally
understandable that, when for any reason there is an unexpected
surge upwards in the demand for a particular resource, leading
to a temporary imbalance in the relationship between supply
and demand, the fear is often expressed that the known reserves
of that resource appear likely to run out in 15 or at the most 20
years. In 1920 the Chief Geologist of the United States' Geological Survey reported that recoverable petroleum still in the
ground amounted to no more than about 7 billion barrels, and
that within five years (and possibly 3) the petroleum production
of the United States would pass its peak. Oil production in the
United States, he was convinced, would cease in the mid 1930s.
The purpose of recording, once again, this notorious misjudgement is not to underline the conservative tendencies of many
geological appraisals, but rather to stress how conceptual inadequacies can severely constrain the understanding of reality.[15]

It is of the utmost importance, therefore, when approaching
the evidence available on energy and mineral reserves, to inter-
pret the data for what it is, and to bear in mind the discounted
value of longer-term reserves to the resource industries.

The third consideration that is basic to an understanding of
resource appraisal follows logically from the above and concerns
the definition of the resource base and its component parts.
Energy raw materials and minerals, it must be recalled, have no
inherent value *in situ*. They are accorded a value by the national
or international politico-economic systems within which they
might be used. The demand for any resource is determined by
a complex set of market, technological and institutional factors.
The supply of a resource is again a compound of, in this case,
geological, technological, transport and political considerations.
Demand and supply are reconciled through allocation processes
which sometimes rely upon market prices, sometimes respond to
an accountant's valuations, sometimes reflect a firm's corporate
strategy and priorities, and sometimes are a function of essen-
tially political considerations.

For example, taconite is a very low grade iron ore containing
between 24 and 29 per cent iron. It is particularly abundant in

Minnesota and Michigan. Historically, it had no value since various grades of direct shipping ore with an iron content of 49–51 per cent were readily available in Mesabi and the other Upper Lakes ranges. In addition, taconite simply could not be used as a blast furnace burden. However, with the development of new beneficiation and agglomeration technologies during the 1950s, and in particular the perfection of pelletizing techniques, taconite rock emerged as an alternative source of iron ore for the iron and steel industry not only in a technical sense but also as a marginally economic reserve. The major factor restraining its increased use in the United States by 1960 was in fact political. This was the uncertainty which executives in the iron ore mining and the iron and steel companies felt concerning the prospective level of corporate taxation in the State of Minnesota in particular. The historical tendency of Minnesota to lean heavily upon the mining industry for incremental revenue raised doubts about the medium term viability of taconite mining. It was therefore only following the voters' approval of an amendment to the State constitution in 1964—the so-called Taconite Amendment, which guaranteed that the rate of tax on iron ore mining companies would not exceed the average level of company taxation in the State—that the major iron ore and steel interests began to invest heavily in the exploitation of the low grade ores.[16] Within a few years, these ores—plus foreign imports—began to displace those of many of the older mines based upon direct shipping ore, partly because the reserves of the latter were running low, but more especially for purely economic reasons. Taconite pellets significantly improve the performance and the economics both of the blast furnace (they are highly reduceable, have a consistent quality and normally contain between 63 and 65 per cent iron) and of the ore transport operations (pellets do not freeze and can be moved throughout the winter months). These advantages more than offset the high costs of agglomeration. The direct shipping ores, on the other hand, not only suffer from the relatively high costs of underground mining in many localities, but they can only be transported seasonally and are not amenable to pelletization. By 1971, in fact, all the underground mines in the Upper Lakes region had been closed down, and over 70 per cent of the ore shipments from that district were in the form of pellets. Without the political guarantee of the Taconite Amendment, however, it is possible that the Minnesota taconites would still remain only a potential ore reserve.

If a mineral deposit or fuel is given a value by the political economy within which it is marketed or produced, it follows that any assessment of the scale of reserves that is available (at present costs and prices, and with contemporary technology) must be similarly qualified. American reserves of iron ore today, for example, can only be realistically measured in relation to present demand and supply technologies and present price levels within the existing economic and political system. Any change in these constraints will inevitably call for a revaluation of the reserves. In a recent study of Californian and Nevada iron ores, for example, it was judged (in the light of both Californian and Japanese market trends and forecasts) that the quantity of 64 per cent iron pellets recoverable from known resources is 594 million tons on the assumption of a price of $15·00 at either San Francisco or Long Beach. However, if the assumed price is raised to $20·00, the reserves stand at 1,321 million tons; and they are estimated to be 1,605 million tons with an assumed f.o.b. export value of $25·00 per ton.[17] Any shifts in the technological or political assumptions underlying these estimates would demand a similar variation in the calculated level of reserves.

The relative nature of resource evaluations in theory is paralleled by the variable degrees of accuracy with which the size of reserves can be measured in practice. Within a given set of socio-economic conditions, some deposits have been surveyed and measured with a fair degree of accuracy. These are the deposits for which detailed geological and technological evidence is available and, if they appear capable of being commercially exploited, they can quite properly be called 'measured reserves'. Extensions of the same deposits, or similar mineral occurances, for which detailed geological and other evidence is not available but about which well-informed assumptions might be made, fall into a separate category of 'inferred reserves'. In addition to the reserves so defined, there will also be quantities of a mineral or fuel that are likely to yield to economic exploitation following some improvement in technology, an upward shift in price, a protective political act, or a change in transport arrangements and costs. These can be defined as the 'potential reserves', some of which are marginal, and others sub-marginal. Together with the very low grade deposits of fuels and minerals, these several types of reserve make up the 'resource base' of the energy and mineral industries.[18] It is imperative to bear these several facets of the resource base in mind in any evaluative exercise, and to be mindful of the fact that knowledge is differentially available—and differentially reliable—in each category.

The importance of these three considerations—the need to focus upon the medium term, the limited quantitative evidence available on energy and mineral reserves, and the several facets of the resource base—can be clarified through the consideration of a single resource. One of the most facinating in the context of the United States is oil. The American economy, it should first be recalled, affords the largest single national market for oil in the world, consuming some 5·5 billion barrels in 1971. The United States domestic industry is also the largest in the world, producing some 3·5 billion barrels in the same year. Any inquiry into the adequacy of the industry's domestic reserves must acknowledge the distinctive institutional and political framework within which it has traditionally operated. The American oil economy, more than that of almost any other resource, has evolved under a unique set of institutional and political constraints. Indeed, conditions during the last 15 years make it more appropriate to think in terms of incentives rather than constraints.

The production phase of the American oil industry, at least in the 'lower' 48 States, is characterized primarily by relatively high costs in comparison with those of other national producers. The various fields and pools naturally exhibit a wide range of costs, the lowest of which are associated with the larger structures of Texas. However, the figures published by Adelman[19] in his study of international oil production costs demonstrates the basic position. Whilst the average operating and development costs of the United States' industry in the period 1960–3 were 122 cents per barrel, five years later comparable average costs

in Venezuela were 46 cents; in Nigeria 16 cents; in Libya and
Saudi Arabia, 16 and 9 cents respectively; in Iraq, 7 cents. It is
small wonder therefore that the average well-head price of
domestic oil in the United States in 1970 was about $1·50 per
barrel more than that of Middle Eastern crude oil landed at
east coast American ports.

The relatively high cost of the United States domestic oil
industry is a function both of the inherent quality of the oilfields
and of the mode of their development. The 'law of capture' and
the desire of many small producers to share in the wealth
created by the recovery of oil, historically produced a much
denser pattern of drilling than was technically necessary to drain
the pools. In eastern Texas, for example, some 25,000 wells were
drilled (mostly during the 1930s) with an average spacing of
one every two hectares. Most expert opinion confirms that this
is ten to fifteen times the number of wells that is technically
required. The wasteful and costly nature of such an institutional
framework is all too apparent. (By contrast, the more recently
developed fields in south Alaska, around Cook Inlet, have been
exploited on a more controlled basis. The average output per
well per day is well over one hundred times the national average, a yield which is able to offset the higher costs of initial
exploration and development there. The lesson of wasteful
reservoir development in the 1930s has been well learned.)
The predictable result of the mode of oilfield development in the
'lower' 48 States, however, was to create a greater level of
productive capacity than the market could absorb. In turn this
began to depress the market price for oil, and the reaction was
the creation of a set of regulatory agencies whose task it has been
to control the level and the geography of American oil production. The principal device to this end has been the prorationing
system, by means of which production quotas are regularly
allocated by the various State regulatory bodies (such as the
Texas Railroad Commission) to individual fields and wells, and
inter-State disagreements are resolved by the Interstate Oil
Compact Commission. For many years, the average American
oil well was grossly underused, pumping oil on only one day in
every three. Only with the energy shortage of the early 1970s,
in fact, was domestic production capacity fully utilized. The

prorationing system was justified on the grounds of conservation, and the need to prevent the wasteful use of an irreplaceable natural resource. But in essence the primary objective of the regulatory institutions has been to maintain the price of oil in the domestic market.[20]

Success to this end has had some important consequences for the behaviour and technology of the American oil industry. Particularly noteworthy is the fact that the relatively high value of oil in the domestic market has stimulated to an unparalleled extent the development of secondary recovery techniques. In most of the world's oilfields only about 30 per cent of the oil in a pool or reservoir is recovered by conventional pumping procedures. In the American case, however, the development of water, gas and steam injection techniques has permitted recovery rates of 40 per cent and over. There are even prospects, with further technological advance (such as the injection of fire and carbon dioxide into the oil bearing structures), of recovery rates being pushed upwards to 60 per cent and more. Whilst oil prices are relatively high, such techniques can be economic as well as feasible. A fall in oil prices on the other hand, would tend to throw American recovery practice back towards the international norm.

The relatively high cost of the American oil industry in its production phase is complemented by the relatively high costs of at least one of its major transport operations. The largest oil-fields are in the south-west of the country; some of the country's largest markets, by contrast, are to be found in the north-east. The cheapest way of reconciling these contrasting geographies of supply and demand is by means of coastal or ocean tanker. However, the movement of oil between American ports comes under the regulation of the 1920 Merchant Marine Act which requires that all coastal trade between American ports be carried in American vessels, built in American yards and manned by American crews. The extremely high capital costs of these vessels by international standards (until the introduction of 43 per cent shipbuilding subsidies in 1972, the price of American tankers was standing at nearly twice that of Japanese and Western European vessels),[21] plus the relatively high costs of American crews (up to four times the cost of maritime labour hired on the world market), together make the coastal movement of oil very expensive by international standards. Delivered to such east coast cities as Boston, New York, Philadelphia or Baltimore from the Gulf Coast, American oil bears freight rates which stand some 60 per cent above what might reasonably have been negotiated on the open market.

The relatively high production and transport costs of American oil could readily be sustained within the context of prorationing through to the early 1950s, in the absence of significant foreign competition. But with the development and the exceedingly low costs of first the Middle Eastern and later the North African oil fields—and in the wake of steadily falling international tanker costs and freights—the traditional political economy of the American oil industry demanded modification. The response of the industry was to seek some degree of protection from overseas competition. When in the 1950s abundant supplies of oil on the world market began to depress prices, and foreign crude and products began to move into the American market on a considerable scale for the first time, the Eisenhower Administration called for a voluntary limitation upon imports. The appeal was barely heard and the predictable failure of this approach to market management led to the imposition of mandatory quotas in 1959. These quotas—which do not apply to the west coast market and which embody certain exceptions such as the 'overland' movement of Canadian and Mexican oil—were originally set at a level equivalent to about 18 per cent of the total national market, or approximately the level of import penetration in 1959. Since that date imports have been allowed to increase slowly to a 1970 level of 23 per cent, and an even higher level of 29 per cent of the domestic market in 1972—first through the decontrol of east coast residual fuel oil imports and then through the 1972 permission to import against 1973 quotas. The clear failure of the quota system to restrain imports led to the 1973 decision to replace it over a seven year period with 'licence fees' for crude and product imports. Along with other measures, such as a tripling of the offshore leasing acreage, this step is designed to keep oil imports, and especially Arab oil imports, down to a minimum.

In brief, these are the major (1973) characteristics of the

economic and institutional framework within which the United States' oil industry continues to operate. And it is within this framework that the industry must necessarily evaluate its reserves in general, and what are known as its 'proved reserves' in particular. Alter any one of these characteristics and the reserve position of the United States' oil industry is likely to change. For example, the abandonment of the mandatory quota system without the imposition of alternative import restraints would undoubtedly have called for a bold downward revision of the level of U.S. domestic oil reserves. On the other hand, the extension of the quotas to the Californian market would have demanded a revision in an upward direction. 'Proved reserves' are defined as those quantities of crude oil that can be extracted by present methods from fields completely developed or sufficiently explored to permit reasonably accurate estimates. They are in essence the minimum reserve figures for the domestic oil industry and in 1970, excluding the North Slope of Alaska, they stood at 37·1 billion barrels of crude oil and natural gas liquids (Table 2). A number of estimates have

Table 2   United States: Production and Proved Reserves of Liquid Hydrocarbons (million barrels)

	Production	Proved Reserves* (end-year)	Ratio of Reserves to Production
1960	2,903	38,429	13·2
1965	3,242	39,376	12·1
1970	4,067	37,104	11·5
1971	4,002	35,767	11·3

* Excluding North Slope of Alaska.
Source: American Petroleum Institute.

also been made of 'oil in place', that is the sum of both reserves and potential reserves. Once again excluding the North Slope of Alaska, these range from 145 to 590 billion barrels and are based upon a considerable variety of hypotheses and premises, some of which are more precise than others.[22]

How adequate are these reserves? One of the more recent authoritative forecasts suggests that oil demands in the United

States will rise from 5·5 billion barrels in 1971 to 9·5 billion barrels in 1985.[23] To meet this demand, 'the physical availability of natural resources in the ground as yet poses no constraints on the use of domestic as opposed to foreign oil'.[24] However, the sole use of domestic oil to meet this level of demand in 1985 would require a quickening of the rate of new oil discoveries, a rate which in recent years has not been sufficient to maintain the reserves to production ratio of a decade ago (Table 2). Whilst the decline in that ratio was only to be expected as criticism of the government's oil import quotas mounted and the threat of increased imports rose, the evidence also suggests that an improvement in the reserves to production ratio will only be achieved at a higher market price than has been paid for oil in recent years. The availability of oil reserves from domestic resources, therefore, will be powerfully conditioned by a combination of market forces and public decisions which together will shape the average level of oil prices, the magnitude of imports and hence the strength of inducements to explore for additional supplies of oil. Public policies also intrude upon this question in a variety of other ways. The level of exploration and drilling activity is strongly influenced by fiscal policies, such as the depletion allowance and other tax provisions. The location of exploitation and drilling activity is especially subject to government influence since a large part of the estimated future discoverable oil in the United States lies under Federal lands, particularly in offshore areas. The medium term supply of oil from domestic sources will undoubtedly be supplemented in the future, as in the past, by domestic condensate and natural gas liquids; here, once again, Federal government policies, which in the past have restrained non-associated gas production rates through low field prices, can play an important role in shaping future developments.

Moreover, other more general aspects of the American energy market are equally subject to public manipulation and so will help to shape the future behaviour of the domestic oil industry. For example, the evolution of natural gas regulation and import controls, the development of attitudes and policies towards environmental pollution, and the extent to which public funds are used to accelerate the arrival of competitive nuclear power will strongly affect the supply, the demand and the allocation of energy in the American economy. All these and other matters bear upon the framework within which oil reserves must be evaluated. Possibly the most central public issue of all concerns the level of oil imports, and the degree to which it is judged acceptable that the American economy should increase its dependence upon foreign sources of such a strategic raw material. With many of the sources of low cost oil characterized by a relatively high degree of political instability, and with the relationship between the international producing companies and the host governments passing through a somewhat fluid stage, such judgements are not readily reached.[25] Certainly the prospect of a continuing rise in the level of United States oil imports, which could well shift the 1972 balance of payments deficit of $4 billion towards $20 billion by 1985, and which could help to create an enormous treasury of Western European and Japanese as well as American funds in the Middle East, cannot be viewed with equanimity.[26]

In other words, questions concerning the adequacy and level of United States' domestic oil reserves elude an absolute answer. Determined by a changing and changeable set of economic and political considerations, the reserves of oil available to the American economy in the medium term are to a considerable degree far from preordained. Rather, they are constrained by a number of key public policies.

This point can be driven home with singular force through a consideration of one of the latest and largest oil discoveries in the country, on the Alaskan North Slope. The first successful oil strike there was in 1968. Estimates of the field's reserves have ranged from 5 billion barrels (at first) to 50 billion barrels at the upper end of an informed but still speculative scale.[27] The field remains relatively unexplored and the amount of evidence from drilling is still limited. Nevertheless, it is clear that the magnitude of the reserves on the North Slope in due course will rest heavily upon the resolution of four uncertainties—the costs of oil production, the expense of oil transport, the trends in national market prices for energy, and a cluster of conservation issues—all of which are subject to varying degrees of public control.

Preliminary estimates of North Slope oil production costs suggest that the expense of drilling will average out at 6 to 7 times the level experienced in California, that is at about $1·5 million per well. At a conservative level of output this can be converted into a cost of $200 per daily barrel; and assuming a decline rate of 5 per cent, plus an opportunity cost of capital of 15 per cent, capital costs can be assumed to stand at about 11·5 cents per barrel. Operating costs, by Adelman's estimates, will be in the order of 6–9 cents per barrel.[28] Total production costs, in other words, are likely to be in the range of 20–25 cents per barrel. This is not particularly cheap by world standards; but the quality of the oil appears to be quite high and it is low in sulphur. To reach a full production cost, a royalty has to be added to this figure.

The transport problems facing North Slope oil are considerable. With the 'Manhattan' trials suspended on grounds of cost, the oil industry will have to use pipelines if it is to get the crude out. The pipeline initially proposed by a consortium of oil companies—the Trans-Alaska Pipeline—would pump oil from Prudhoe Bay to the Pacific Coast at Valdez, a distance of some 1,280 kilometres. (An alternative, and economically more attractive, proposal to pump oil directly to the markets of the Middle West via the Canadian Mackenzie Valley has not been received with notable enthusiasm by the oil companies.[29]) The huge 22 centimetre pipes for the Trans-Alaska Pipeline were delivered to Valdez in 1969. Its construction has been expensively delayed for several years, however, by native land claims and litigation under the Natonal Environmental Protection Act. Conservationists have raised objections to the pipeline on the grounds that its construction would cause a serious physical disturbance in the country's largest remaining wilderness area, that pipeline leaks or breaks could be occasioned by melting permafrost, seismic disturbances or avalanches, and that the migratory habits of the caribou herds that inhabit the area would be adversely affected. By 1973, after more than three years of debate and litigation it appeared likely that a special Act of Congress would in fact be passed to ensure the construction of the pipeline. Assuming that safeguards to the ecology, plus the costs of delay, do not more than treble earlier estimates of the capital cost of the pipeline, the unit costs of moving the crude oil to Valdez appear likely to be in the order of $1·00 per barrel. This would yield an approximate cost of North Slope oil f.o.b. Valdez of $1·25 per barrel, plus royalty.

The expense of transporting the oil to the Californian market—say, Long Beach—poses further uncertainties. Assuming that the best international maritime practice could be adopted —that is, large supertankers of 150,000 dwt.–200,000 dwt., chartered on a long-term contract of affreightment—the freight rate could be as low as 12–15 cents per barrel. However, quite apart from the fact that no Californian port can as yet handle vessels in excess of 150,000 dwt., it has to be recalled that the Mercantile Marine Act would generate a surcharge on world rates. Before the recent introduction of shipbuilding subsidies through the Maritime Administration, this surcharge would have been about 60 per cent on 'world' rates; prospectively it might be reduced to about 30 per cent. The cost of moving oil from Valdez to Long Beach, therefore, appears likely to cost approximately 25 cents per barrel and so to help produce a total delivered cost of North Slope crude oil in the Los Angeles market of about $1·50 per barrel (plus royalty). The profitability of selling the oil in the Californian market is unquestioned. Posted prices in the Los Angeles region in recent years have ranged between $3.25 and $3·55 per barrel for 36° gravity crude, and under $3·00 for the lighter 28° supplies—an average price of about $3·00 per barrel. There would also be clear profit in shipping the oil via the Panama Canal to the east coast market; for in recent years oil has been valued there at approximately $3·50 per barrel. With Kuwait supplies being landed in Japan at just under $1·50 per barrel in 1970, the flow of Alaskan crude to Japan, despite the probability of lower freight rates than those applicable to the southern California run, would be likely to show a somewhat smaller profit.

Whilst the United States market for oil remains characterized by relatively high prices, it is clear that Alaskan North Slope oil could bear a considerable increase in its production and transport costs and still remain highly profitable. Consequently, the incentive to explore for more oil, and the prospects for the discovery of additional reserves there, are very much greater than if lower oil prices were to be encouraged by public policies in the United States. However, the existence there of anything more than potential reserves rests upon the assumption that the objections of the Alaskan Eskimoes and the conservation lobby can be either met—or overruled. Without the necessary political decisions, North Slope oil can make no contribution at all to the oil reserves of the American economy.

A consideration of the specific instance of the Alaskan oil field, and the more general case of the total oil wealth of the United States, illustrates the difficulties involved in providing simple answers to even the broadest questions of resource adequacy. It underlines, however, the need to recognize three principles in any interpretation of America's resource position. First, natural resources must be seen as a relative and not an absolute endowment. To echo a former President of Standard Oil of New Jersey, 'Raw materials do not exist; they are created'. Second, with the data base for most resource appraisals so imperfect, evaluations can be no more than provisional and must be revised in the light of constantly changing evidence. Third, and perhaps most important, it is imperative to acknowledge that the resource future of the American economy is not pre-ordained by natural endowments; rather it will respond to political, politico-economic and corporate decisions, and the implications of technological advance, as much as to the distinctive geological attributes of the country. The energy shortages of the early 1970s are a measure of the imperfections of the political economy of the United States rather than a reflection of resource inadequacy in an absolute sense. By the same token, any problems of resource supply for the United States throughout the rest of the 1970s are more likely to stem from failings in the national and international political economy rather than from the natural endowments of the country.

To respond more directly to the question posed in the title of this essay—is the cornucopia vanishing?—an answer can be framed in the following terms. As a consequence of the steady increase in the demand for fuels and mineral raw materials, and also in response to a general preference for the availability of these materials at a reasonably low price, there is a medium term tendency for the United States to become increasingly dependent upon overseas supplies. But while international trade and comparative advantage currently encourage this characteristic in general, with the exception of a few relatively rare minerals, there are no resource, economic or political reasons why the trend should of necessity continue at any predetermined rate. The resource options of the American economy, in other words, remain remarkably wide open. Each option of course has a different total cost and is associated with a different set of risks. The fundamental resource problem of the United States, therefore, concerns the proper evaluation of the many alternatives, the choice of a policy option and the creation of appropriate mechanisms that will ensure its effective implementation.

NOTES

1. Bureau of Mines, *Minerals Yearbook 1970*, 1, Govt. Printing Office, Washington D.C., 1972, p. 332. This figure includes semi-bituminous coal, lignite and anthracite reserves which comprise rather more than half the total.

2. J. L. Fisher and N. Potter, *World Prospects for Natural Resources: Some Projections of Demand and Indicators of Supply to the Year 2000*, Johns Hopkins Press for Resources for the Future Inc., Baltimore, 1964, p. 14.

3. President's Materials Policy Commission, *Resources for Freedom*, Govt. Printing Office, Washington D.C., 1952.

4. H. H. Landsberg, L. L. Fischman and J. L. Fisher, *Resources in America's Future*, Johns Hopkins Press for Resources for the Future Inc., Baltimore, 1963.

5. Bureau of Mines, op. cit., p. 912.

6. Cabinet Task Force on Oil Import Control, *The Oil Import Question*, Govt. Printing Office, Washington D.C., 1970, p. 124.

7. R. and E. Ehrlich, *Population, Resources, Environment*, W. H. Freeman, San Francisco, 2nd edn., 1972; J. W. Forrester, *World Dynamics*, Wright-Allan Press, Cambridge, Mass., 1971.

8. See, for example, U.S. Department of the Interior, *United States Energy: A Summary Review*, Govt. Printing Office, Washington D.C., 1972.

9. See, for example, National Petroleum Council, *U.S. Energy Outlook: An Initial Appraisal 1971–1985*, NPC, Washington D.C., 1971.

10. H. H. Landsberg *et al.*, op. cit.

11. American Petroleum Institute, *The Energy Supply Problem*, API, Washington D.C., 1970.

12. M. A. Adelman, *The World Petroleum Market*, Johns Hopkins Press for Resources for the Future Inc., Baltimore, 1972.

13. M. K. Hubbert, 'Energy Resources' in National Academy of Sciences and National Research Council, Committee on Resources and Man, *Resources and Man*, W. H. Freeman, San Francisco, 1969, p. 173.

14. H. H. Landsberg and S. H. Schurr, *Energy in the United States: Sources, Uses and Policy Issues*, Random House, New York, 1968, p. 98.

15. Ibid., p. 98.

16. Gerald Manners, *The Changing World Market for Iron Ore, 1950–1980: An Economic Geography*, Johns Hopkins Press for Resources for the Future, Inc., Baltimore, 1971, p. 155; Gerald Manners, *The Changing World Market for Iron Ore: A Descriptive Supplement Covering the Years 1950–1965*, University Microfilms for Resources for the Future Inc., Ann Arbor, Michigan, 1971, pp. 2–8.

17. L. Moore, *Economic Evaluation of California–Nevada Iron Resources and Iron Ore Markets*, U.S. Dept. of Interior, Bureau of Mines, Govt. Printing Office, Washington D.C., 1971, p. 58.

18. Gerald Manners, 'New Resource Evaluations' in R. C. Cooke and J. H. Johnson, *Trends in Geography*, Pergamon, London, 1969.

19. M. A. Adelman, op. cit., p. 76.

20. W. F. Lovejoy and P. T. Homan, *Economic Aspects of Oil Conservation Regulation*, Johns Hopkins Press for Resources for the Future Inc., Baltimore, 1967.

21. *The Economist*, 5 August 1972, p. 41.

22. Hubbert, op. cit.

23. National Petroleum Council, op. cit., p. 28.

24. Ibid., p. 24.

25. S. H. Schurr and P. Homan, *Middle Eastern Oil and the Western World, Prospects and Problems*, American Elsevier Publishing, New York, 1971.

26. J. J. Murphy (ed.), *Energy and Public Policy, 1972*, Conference Board, New York, 1972.

27. Ibid., p. 86.

28. M. A. Adelman, *Significance of Shifts in World Oil Supplies*, paper presented to 20th Alaska Science Conference, Fairbanks, 24–7 August, 1969.

29. C. C. Cicchetti, *Alaskan Oil Alternative Routes and Markets*, John Hopkins Press for Resources for the Future Inc., Baltimore, 1972.

First drafted in 1971, this essay has had to ride on the rising tide of what was first an American, and then subsequently a global, 'energy crisis'. As a final footnote, dated January 1974, therefore, it should be recorded that, following the Arab oil production cutback and embargo upon oil exports to the United States in October 1973, world energy markets were thrown into considerable disarray and the existing price structure for oil quickly disintegrated. In response American policy, coordinated through a new Federal Energy Office, moved positively in the direction of ensuring that the country would have a high level of energy self-sufficiency in the medium term. A Federal right of way permit was given for the Trans-Alaska Pipeline in January 1974.